SEA STINGERS – AND OTHER VENOMOUS AND POISONOUS MARINE INVERTEBRATES OF WESTERN AUSTRALIA

SEA STINGERS

– AND OTHER VENOMOUS AND POISONOUS MARINE INVERTEBRATES OF WESTERN AUSTRALIA

by

Loisette M. Marsh, Dept of Marine Invertebrates
Shirley M. Slack-Smith, Dept of Malacology
Western Australian Museum

with medical contributions by

Desmond L. Gurry, Senior Lecturer in Child Health
University of Western Australia

With best wishes
Loisette.

Western Australian Museum 1986

ISBN 0-7309-0326-5

Cover: Northern Blue-ringed Octopus
Portuguese Man-of-War
Needle-spined Urchin

Cover design: Jill Yelland

Published by the Western Australian Museum, Francis Street, Perth, Western Australia 6000.

Printed for the Western Australian Museum by Dix Print Pty Ltd, Perth, Western Australia.

museum

Foreword

It is a delight to explore the sea-shore and the shallow offshore waters with their great diversity of animal and plant life, but the pleasure can very quickly change to pain if sensible precautions are not taken when handling some of the beautiful and seemingly harmless animals that are found there.

This handbook sits well in the tradition established by the Western Australian Museum of providing the public with informative and readable accounts of aspects of the local fauna. The text enables readers to recognise potentially dangerous animals, gives interesting and useful information about them and their venoms, and reactions to the many and varied poisons they produce. There are other books on such animals in Australia, but it is the authors' practical experience and their local knowledge of the animals and where they are found that makes this book particularly valuable.

Museum biologists Loisette Marsh and Shirley Slack-Smith have both dived widely, collecting and studying the invertebrates of Australian and Indo-Pacific waters, and they have been able to provide new information on some of the stingers and their effects on their victims. Dr Desmond Gurry has a deep interest in the medical aspects of this field. While pointing out the limitations to present knowledge of the great diversity of poisons and the way to combat them, the book succinctly gives the best advice currently available on treatment.

It is fascinating to learn of the many ingenious devices that animals have evolved for defensive or offensive purposes, from poison laden harpoons to the second-hand use of the defensive weapons of their prey, weapons they manage to neutralise and employ themselves for defense. While the main purpose of this book is to help people to recognise the venomous animals of our shores and to relieve injuries when they occur, it is also to be welcomed for much else in it that should stimulate an interest in the life of our Western Australian shores.

(ERNEST P. HODGKIN)

Formerly Trustee, and Chairman of Trustees, of the Western Australian Museum

hydroid
Portuguese man-of-war
saucer jellyfish
sponge
Chironex box jellyfish
bristle worm
box jellyfish
xanthid crab
sea anemone
coral
brittle star
starfish
nudibranchs
sea cucumber
sea squirt
sea urchin
octopus
mussel
cone shell

Contents

Illustrations

Photographic Credits

C.C. Appleton: Figures 39, 41
Patrick Baker: Figures 27, 28, 44
Isobel Bennett and Keith Gillett: Figures 14, 17
Clayton Bryce: Figures 6, 8, 9, 12, 26, 29, 30, 32, 34, 37, 42, 46, 47, 48, 50, 52, 54, 55, 56, 57, 58, 59, 60, 61, 62, 64, 65, 67, 68, 69, 70, 71, 72, 75, 82, 83, 86
Neville Coleman: Figures 11, 33, 35
R.J. Heazlewood: Figure 23
J. Barry Hutchins: Figures 77, 84
Loisette Marsh: Figures 76, 80, 87
Roche Products Pty Ltd: Figures 5a, b, 21. Reproduced with kind permission.
Geoffrey Shaw: Figure 10
R.V. Southcott: Figure 2
J. Williamson: Figures 18, 20, 22
Barry Wilson: Figures 7, 12, 36, 38, 43, 45, 49, 73, 74, 78, 89

Diagrams — drawn by Jill Ruse

Figure 1a – redrawn from Bergquist 1978
1b – redrawn from Branch 1981
3 – redrawn from Branch 1981
4 – adapted from Branch 1981 and Cleland and Southcott 1965
13 – redrawn from Branch 1981
15 – redrawn from Southcott 1967
16 – drawn from photograph in Halstead 1965
19a – drawn from Barnes 1966
19b and d – drawn from photographs by Barnes 1966
24 – redrawn from Branch 1981
25 – original drawing, L. Marsh and J. Ruse
31 – redrawn from Barnes 1980
40 – adapted from Appleton 1980
51 – original drawing S. Slack-Smith and J. Ruse
53 – original drawing S. Slack-Smith and J. Ruse
63 – original drawing S. Slack-Smith and J. Ruse
79 – redrawn from Buchsbaum 1951
81 – redrawn from Mortensen 1935, 1940
85 – redrawn from Mortensen 1943
90 – redrawn from Branch 1981

Acknowledgements

We thank all those who have so generously lent slides or photographs (listed separately); Dr C.C. Appleton who made available his Ph.D. Thesis on *Austrobilharzia* through Murdoch University; Dr R.W. George, formerly of the Western Australian Museum, for his contribution on toxic crabs; Mr W.H. Butler, Conservation Consultant, for a first hand account of a stinging by *Echinothrix calamaris;* Dr R. Endean, University of Queensland, for providing the case history of stinging by *Tamoya* sp. at Exmouth; Dr B.C. Russell, Curator of Fishes, Northern Territory Museum, for a first hand account of stinging by *Carybdea alata;* Ms M. Stanbury, Western Australian Maritime Museum, Fremantle, for a first hand account of stinging by carybdeid jellyfish at the Rowley Shoals; Dr R. Lethbridge, Murdoch University for critically reading the manuscript; Mr G.W. Kendrick and Mrs A. Brearley for first hand accounts of the results of stinging by *Conus anemone;* Mr U. Weinreich for information on the natural history of *Conus geographus;* Mr and Mrs B. Sharpe for information on the natural history of blue-ringed octopus; Dr Seddon Bennington for adapting the manuscript to a 'popular' format; Mrs Ann Ousey, Mr Greg Jackson and Mrs Shanthi Hildebrandt, the production team of the Museum's Publication Department; and Jill Ruse for the production of meticulous diagrams.

Introduction

The long coastline of Western Australia, extending for over 12,000 km from the tropical north coast to temperate waters of the south coast, attracts thousands of swimmers, skindivers, fishermen, naturalists and casual reef explorers. Others avoid the sea, often through fear of its unknown and presumed dangerous inhabitants.

This handbook gives information on how to recognise and avoid the potentially harmful marine invertebrates (animals without backbones) living in the coastal waters of the region. It also describes the symptoms, first aid and follow-up treatment of stings and envenomation that may result from encounters with these animals. Venomous marine vertebrate animals are described in 'Dangerous Fishes of Western Australia' and 'Dangerous Snakes of Western Australia' published by the Western Australian Museum.

Many marine invertebrates use venoms and toxins for catching prey or defending themselves from predators. Some of the venoms and toxins are potent enough to affect humans while a few are very dangerous. The box jellyfish *Chironex* can kill a child in a few minutes. As well as the dangerous species many other less venomous, but nevertheless troublesome, animals are included. In many cases identification of the culprit and prompt and appropriate first aid can prevent prolonged pain and discomfort or even death. Most of the dangers can be avoided by sensible precautions such as wearing protective footwear and gloves when reefwalking and a wetsuit when diving.

The authors have each spent many years working and relaxing in the sea around tropical and temperate reefs and shores without suffering more than minor injuries. It is hoped that this handbook will help many others to enjoy our coastal waters in safety.

Sponges

Phylum Porifera

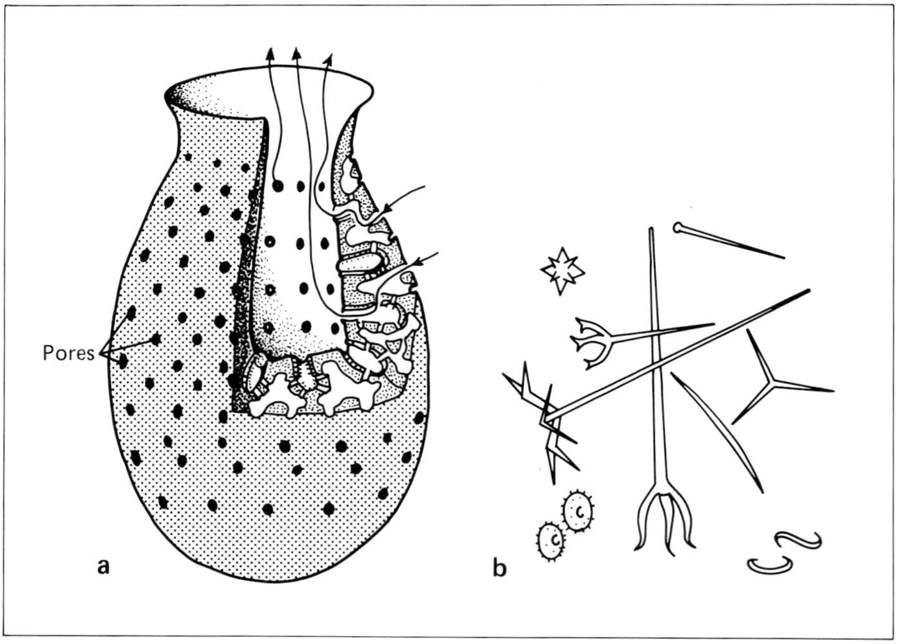

Figure 1: (a) A simple sponge cut to show the interconnecting canals which carry water through it in the direction arrowed. (b) Some examples of sponge spicules.

Sponges despite their plant-like growth are animals. They are colonies of semi-independent cells that feed by straining minute organisms from the stream of water that enters through pores all over the body and leaves from one of the larger holes at the top (Figure 1). These millions of pores give the sponges their scientific name Porifera or 'pore-bearers'.

Most sponges live in the sea, and range from those which have a definite shape to others which are encrustations on rocks. Shallow water sponges are often brightly coloured, particularly the red, green, yellow, orange, pink or mauve sponges found under boulders or growing on the walls of undersea caves. Deeper water sponges are more often dull coloured or white. They range in size from as small as a thumbnail, to a metre or more in height. The sponge skeleton is a meshwork of horny spongin fibres strengthened in most by glassy spicules (or needles) of silica. Soft bath sponges have a skeleton of spongin, without spicules.

Sponges contain a variety of toxic compounds to deter potential predators and the secretions from some cause severe skin rashes if touched.

A toxic sponge *Neofibularia mordens* (Figure 2) has been found in South Australia and is also likely to occur on the south coast of Western Australia.

It is a massive sponge, as much as 40 cm high and wide. Alive it is royal blue with a smooth slimy surface but out of the water it turns to a dark chocolate bluish-brown colour. The cut surface is light brown with a gritty texture. Contact with the slime causes intense itching.

Several other species in southern and northern Australia are known to cause skin rashes but they are not well documented.

Sponges, alive or dead, cause irritation in different ways, each requiring different treatment:

The needle-like glassy spicules can pierce the skin when the sponge is handled, or can enter the eyes of seamen as they haul up deep-water trawl nets. The spicules can also be inhaled or swallowed, particularly after the sponge fragments have dried out.

Toxin in the slime of some kinds of sponges causes severely painful contact dermatitis. The chemicals may also be damaging to the eyes.

Creatures living on and in sponges (e.g. anemones and bacteria) may cause infections.

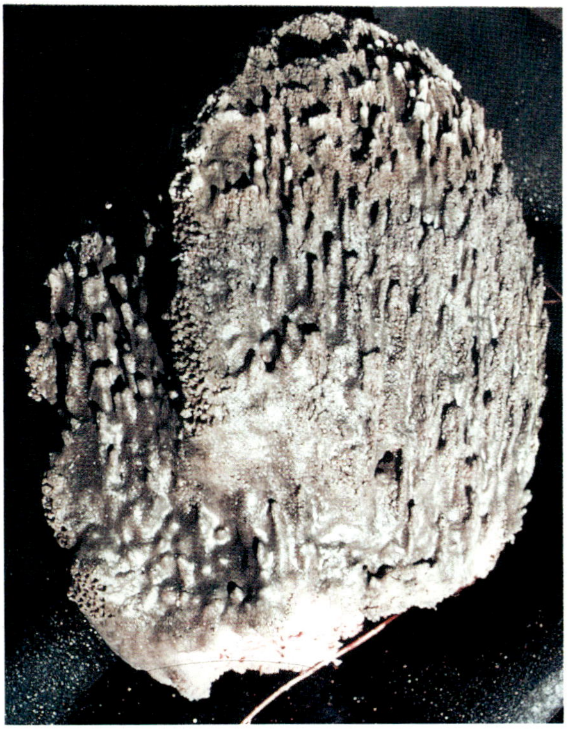

Figure 2: A southern Australian sponge *Neofibularia mordens* which causes persistent skin rashes when handled.

Injury Prevention

- Never handle a sponge, dead or alive, without gloves.
- Avoid touching or sitting where sponges have been lying.
- Use safety glasses when emptying trawl nets.

Symptoms

- These consist of itching, burning and sometimes redness.
- There are cases of intense itching pain often starting hours after touching a toxic sponge and lasting for days, which is resistant to all attempts at relief. There may be little to see in the skin which is hurting so much. In other cases an obvious itching, swelling, redness of the skin (urticaria) may develop. Symptoms might last for weeks.
- The itching has been described by the victims as like having glass in the skin!

Treatment

- Visible glass spicules should be removed. An attempt can be made to remove invisible glass spicules by applying adhesive tape and pulling it off.
- There is no clear evidence that any known treatment is of great value. Anti-itching creams and oral anti-histamines have been tried, and even cortisone, by cream and orally. Some individual reports describe relief by things which cool, and sometimes a recurrence of the itching pain when the area is warmed. Frankly the long duration of the symptoms has provided a great opportunity to show the ineffectiveness of most of these suggested remedies.
- The application of vinegar gives relief in some cases.
- Eventually the skin peels off at about the time the symptoms go.
- If material is breathed in and lung irritation appears (with difficulty in breathing, wheezing and coughing perhaps of blood) medical treatment will be directed at those measures used in asthma and against allergy. If the material is swallowed there could be bleeding, for which there is little direct treatment. Aspirin should not be given to anyone who has a bleeding problem.

Jellyfish, Sea Anemones, Corals and Their Relatives

Phylum Cnidaria

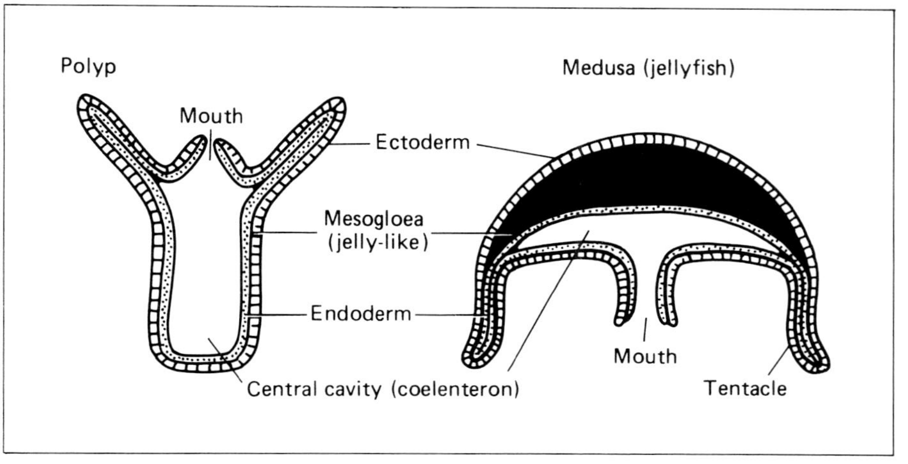

Figure 3: A polyp and a medusa – variations on a theme.

Cnidaria are often plant-like in their growth but they are really simple animals with a body wall surrounding a central cavity which has a single aperture, the mouth, often surrounded by tentacles. The body wall has a gelatinous middle layer that makes up the bulk of the body in such animals as jellyfishes.

There are four main divisions of the phylum, the classes Hydrozoa (sea ferns, hydromedusae, fire corals and Portuguese men-of-war), Cubozoa (box jellyfish), Scyphozoa (jellyfishes) and Anthozoa (sea anemones, corals and soft corals).

Some of these animals live singly but many form colonies, often of complex structure, such as a Portuguese man-of-war or a coral colony.

All are variations on two body forms – the polyp and the medusa (jellyfish) (Figure 3). Hydrozoans usually have a life cycle including small medusae which produce fertilised eggs which grow into sedentary polyps that reproduce asexually to form colonies that are often plant-like in appearance. In Cubozoans and Scyphozoans the polyp is very small and the medusa is the conspicuous stage of the life cycle. Anthozoans form more complex polyps which reproduce both asexually and sexually without a medusa phase.

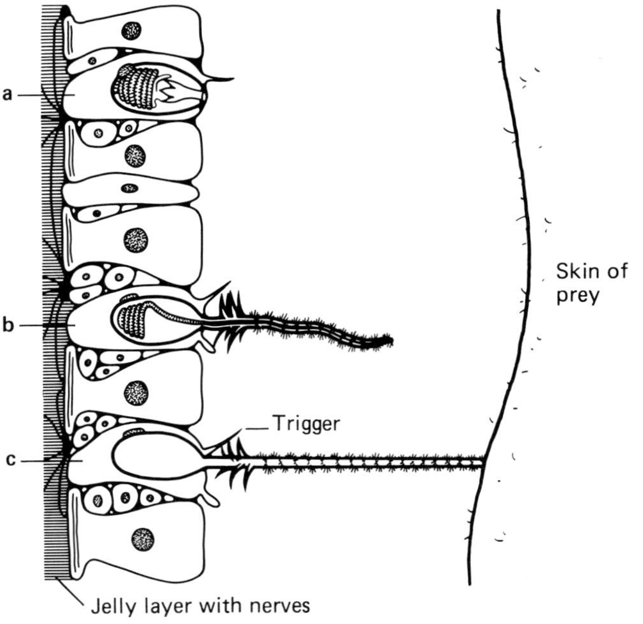

Figure 4: The outer cell layer of a jellyfish tentacle with an undischarged nematocyst (a), one partly discharged (b) and a fully discharged nematocyst (c) with its hollow barbed thread, through which venom is forced into the prey.

The Stinging Mechanism

Cnidarians catch their prey and defend themselves by discharging stinging capsules or nematocysts on the tentacles. The nematocysts are so minute that each one lies within a cell, but groups of them are often visible as beads or rings on the tentacles of jellyfish or Portuguese men-of-war. There may be up to 1,500 nematocysts to a square millimetre in the deadly box jellyfish, *Chironex*. Each minute stinging capsule contains a coiled thread which shoots out by reflex action when its trigger is touched or stimulated by certain chemicals. The thread pops out like the long finger of a rubber glove being turned inside out, and it may stick to the prey, or inject venom. Some nematocysts have closed tubes used for entangling the prey and anchoring it while others inject venom through their open ends. As the prey struggles, more capsules are triggered to discharge and the prey is paralysed by the venom (Figure 4).

Each species of Cnidarian has several characteristic types of nematocyst so that stinging capsules recovered from the skin of a patient can help to identify the stinger. Their shape, the length and arrangement of spines on the threads, the type and number and pattern of stings are also useful guides to the identity of the stinger (Figure 5, a, b).

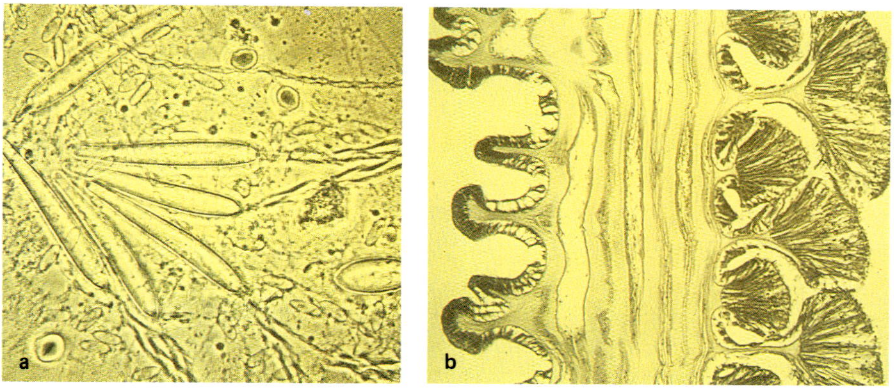

Figure 5: (a) Nematocysts of the deadly *Chironex* box jellyfish. (b) Microscopic section of *Chironex* tentacle with nematocysts.

Most Cnidarians merely feel sticky to touch but nematocysts of some can penetrate human skin. The venom injected by some Cnidarians can spread rapidly and with great effect not only to the region touched but also to distant parts and even to the whole body.

Medical Aspects of Nematocysts

The nematocysts or stinging capsules contain a chemical designed to kill or stun prey. The usual prey is small, even microscopic. The chemicals are fast-acting poisons; they may paralyse or corrode or both. When injected into humans – usually by accident, because these small Cnidarians cannot eat us – they produce inflammation, pain and sickness.

The effect on the human and the severity of the sting will depend on many factors, the nature and strength of the poison related to the type and size of the animal; the concentration of stinging capsules on the animal; the surface area of contact; the time of contact and the resistance of the skin of the human. A small version of a midly venomous jellyfish brushing over the thick skin of the hand of a swimmer may cause a barely noticeable effect. A large jellyfish of dangerous type, hungry and laden with undischarged stinging capsules, caught up and stuck to a large area of soft armpit or torso skin of a human, especially being trapped there by clothing, can cause very severe skin damage and pain and make the victim feel generally very ill.

Remedies – Why Are There So Many?

Treatment must be on three fronts – neutralising venom, relieving pain, and inactivating nematocysts.

Consider a sting of mild to moderate severity. Some stinging threads have injected their toxins into the skin, shallowly. The toxins damage flesh, and cause pain, inflammation, redness and swelling.

The pain comes from the poison already received, and can be added to by other stinging capsules which discharge later. It may be obvious that undischarged or only partly discharged nematocysts are there in an adherent piece of tentacle. Smaller microscopic fragments of tentacles may be present but not so obvious. Studies by the Royal Australian Navy indicate that, at least with the Portuguese man-of-war, continuing pain occurs because there is continuing stinging by active stinging capsules, so measures to stop further stinging are necessary.

The next point is that the pain from the poison already injected by the nematocysts will usually stop after a time. This has caused remarkable confusion and a welter of cures – all of which will seem to work if given enough time. Some are home remedies, some are commercial. Some result from a need for people to feel that they are doing something, and it is hard to sort out the wheat from the chaff. The last chapter on this is yet to be written, and a deep sense of humility is necessary. Perhaps the common remedies of sodium bicarbonate (baking soda) or sodium borate (as in washing blue) will be shown to be of value eventually. Northern Aborigines use plant substances such as the latex-like sap of the *Euphorbia* plant.

They might yet be shown to have real pain-relieving properties.

- Unexploded stinging cells. - Chemical treatments.

For years doctors recommended, and life savers used, alcohol such as methylated spirits 'to kill the stinging cells'. Scientists in Queensland then found that alcohol caused the 'loaded' stinging capsules to fire off better than ever. There was a disconcerting time lapse before official recommendations were changed to use acetic acid in strengths of from 3-10% (i.e. vinegar which contains acetic acid at 4-5%). This acid very definitely inactivates the unexploded stinging capsules in a few minutes. We now wonder why alcohol was used so confidently for so long, and can only reflect that it would make a tentacle visibly shrink.

The microscope further shows that some home remedies such as sea water and urine do not protect against more discharge, but perhaps the actual washing or splashing does remove microscopic pieces of tentacle. Fresh water stimulates the discharge of more stinging capsules so should never be used. Some commercial remedies fail, but some succeed, including 20% aluminium sulphate which is the basis of the highly publicised product, 'Stingose'.

- Physical treatments.

Removing obvious tentacles helps. Rubbing with wet sand has never been in favour, as it causes further discharge of stinging capsules and damages skin already hurt. Throwing hot dry sand (an Aboriginal treatment) or powder of any kind or pouring on oil can prevent more damage. Tight bandaging (as is advised for snake bite) could make skin damage and pain worse.

• Relief of pain from toxin already injected.

This is very hard to assess, because of the self-curing aspect, and other confusing features.

It may be that the methylated spirits remedy of yesterday actually helped by its physical cooling effect, or by acting as a counter irritation. It may be that some chemicals applied on the skin will inactivate the poison there –this claim has been made for aluminium sulphate and also for steak tenderiser which is papain, an enzyme that breaks down protein and comes from the pawpaw. In the Philippines a folk remedy for jellyfish stings is the juice of green pawpaw.

Most marine poisons are destroyed by heat. Heat however can hurt the skin. Inflammation itself can be reduced by cool packs as are used in sporting injuries. Clearly one can not use both these methods at once. We recommend cold to reduce local chemical reactions and the redness and pain. Care must be taken that the cold is not itself damaging.

Local anaesthetic creams have been used, but their action has not been assessed carefully. The pain usually goes away but this might have happened anyway. Perhaps much more concentrated anaesthetic creams will be shown to be more useful than those used currently.

General analgesics or pain relievers such as aspirin, paracetamol, codeine, etc. may help. The pain of more serious envenomations is another story. Treatment involves the use of narcotics, even intravenously. Specific antivenom has been developed against only one jellyfish – *Chironex*.

Part of the effect of a sting may be due to allergy or hypersensitivity. If so, oral antihistamines may help. They are prescribed for patients known to be very sensitive and where weals and itching feature.

Summary of Current Recommendations for Mild Stings

• Stop further stinging by removing or inactivating stinging cells – wash with large quantities of diluted acetic acid (vinegar) or aluminium sulphate (Stingose).

• Remove obviously sticky pieces of animal.

• Only if acetic acid or aluminium sulphate is not available – use dry powder e.g. sand to separate pieces of tentacle from the skin.

• Don't rub the skin. Keep the area rested, and if possible elevated.

• Use cold to reduce inflammation and relieve pain – crushed ice in a plastic bag can safely be applied for periods up to 30 minutes, so can cool cans. Solid ice and especially solidly frozen gel should be used with layers of dry towel, or the cooling can be too severe.

• Local anaesthetic cream and/or Papain could be used.

• General analgesics – aspirin, paracetamol, codeine should help.

• Oral antihistamine might be indicated if there seems to be a strong allergic element.

• For severe stings see page 32 under *Chironex*.

Hydroids

STINGING HYDROIDS

Class Hydrozoa
Order Hydroida

Figures 6, 7: Two species of 'fireweed' or stinging hydroid, *Aglaophenia* (Figure 6) and *Lytocarpus* (Figure 7).

Hydroids are delicate fern-like colonies of animals, from a few millimetres to 20 centimetres in length, living on rocks and jetty piles. Each tiny branch is lined with minute individual members of the colony called polyps, each bearing tentacles armed with stinging capsules or nematocysts which are used for food gathering.

Many hydroids have nematocysts capable of penetrating human skin and which sting to some degree, but two kinds known as stinging or fire-weeds, which are usually associated with coral reefs, do sting painfully. Both occur in north-western Australia and are occasionally found in shallow water as far south as Fremantle. The feather hydroid or stinging 'seaweed' *Aglaophenia cupressina* (Figure 6), which resembles a brown seaweed, grows in clumps with fronds about 15 cm long.

Fire-weed, *Lytocarpus philippinus* (Figure 7) resembles a small white bush of delicate fronds with many branches arising from a single horny base. It is conspicuous on coral reefs, particularly along reef edges and channels.

Injury Prevention

- Wear gloves and protective clothing while diving or snorkelling.

Symptoms

- From a mild stinging sensation to extreme pain, usually increasing over the first 10 minutes. Pinpoint lesions may follow, developing into a generalised blotchy red rash, with swelling and sometimes blister formation. The rash may last up to 10 days with local discomfort and itching.

Generalised reactions are uncommon, but may include abdominal pain, cramps, nausea and diarrhoea and a general feeling of illness.

Treatment

- Weak acetic acid (vinegar) or aluminium sulphate could be used to inactivate unexploded nematocysts and so to prevent further injury. Topical local anaesthetic creams could be tried though they might produce a skin sensitivity of their own after several days use. Steroid creams could also be tried.

MEDUSAE OF HYDROID POLYPS

Class Hydrozoa

Some small medusae, resembling jellyfish, are the sexual stages of hydroid polyps. Hydromedusae all have four canals forming a cross on the umbrella and are mild stingers.

Order Anthomedusae

Leuckartiara gardineri

The 'little red stinger' *(Leuckartiara gardineri)* is a small, conical medusa only 10 mm high by 5 mm wide. Gonads and the large folded mouth fill the bell. There are four pink marginal tentacles about 25 mm long and numerous minute thin tentacles.

Since it is known from the Indian Ocean and north Queensland *L. gardineri* could be expected on the Western Australian coast, particularly in the north.

The relatively harmless *Leuckartiara* could be confused with the more dangerous Irukandji stinger (page 22) which is similar in size but has a warty bell with the four tentacles borne on pedalia (Figure 15) and no marginal tentacles.

Symptoms

- A sharp sting that rapidly subsides.

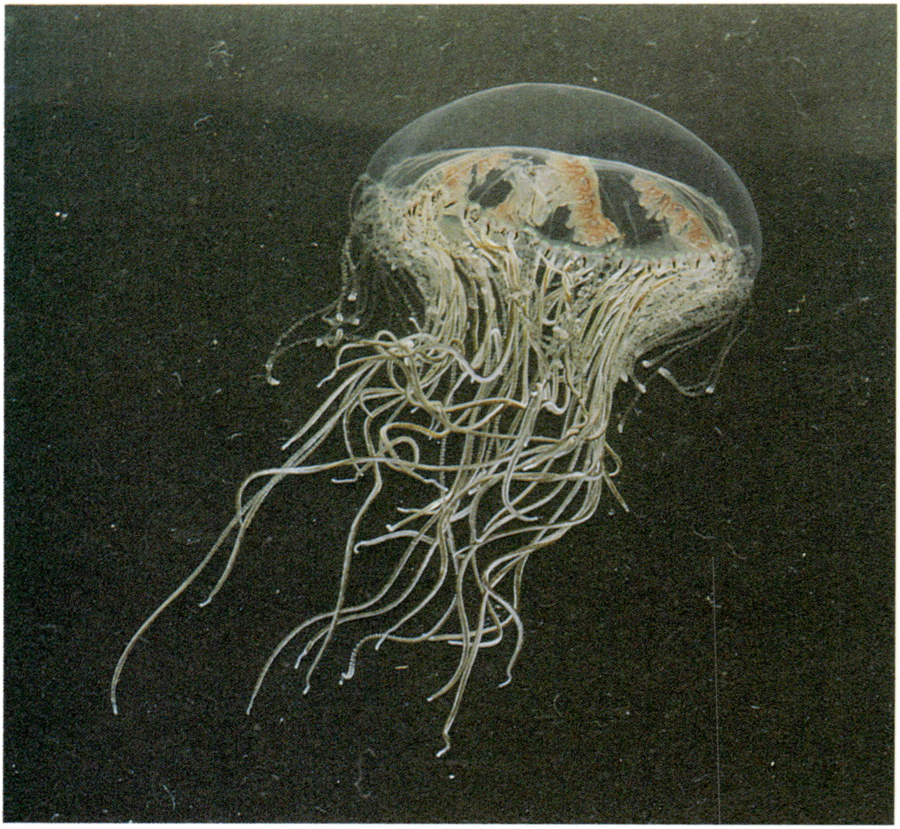

Figure 8: Olindias.

Order Limnomedusae

Olindias singularis

The small, hemispherical medusa of *Olindias singularis* (Figure 8) is clear, firm and 1-3.5 cm across. It has a central mouth and four radial canals, each with a frilly pink gonad along part of its length, and several short unbranched canals running inwards from the margin. It has a fringe of short marginal tentacles and many longer tentacles which, when contracted, are marked with brown beads or rings of nematocysts. The medusa is the sexual stage of a small polyp which lives attached to the bottom.

Olindias singularis occurs in the tropical Indian Ocean and off Queensland and South Australia. In Western Australia it has been found in Shark Bay and near Albany.

Symptoms

- A mild sting, which lasts about 30 minutes.
- A faint pink weal forms after about 15 minutes.

Portuguese Man-of-War

Class Hydrozoa
Order Siphonophora

Figure 9: The blue-bottle or Portuguese man-of-war, *Physalia physalis,* has a gas-filled float which supports the feeding tentacles and polyps below.

Physalia physalis (Figure 9)

This floating colony of medusae and polyps is also known as the blue-bottle because of its conspicuous blue, gas-filled float, which allows it to sail the seas, dragging feeding tentacles and polyps below. Fleets of blue-bottles are carried by warm currents to temperate coasts where they are washed ashore in thousands by storms or onshore winds. In south-western Australia they are generally seen in autumn and winter, less often in summer.

The float is from a few centimetres to more than 15 cm long, and the main tentacle may be 10 metres long when fully extended, or contracted to a few centimetres in length. The tentacle has a frilled and beaded appearance due to the groups of stinging capsules or nematocysts spaced closely along its length. Each bead carries hundreds of nematocysts so that the whole tentacle carries a powerful armament.

When the nematocysts are triggered to explode the tubes uncoil with enough force to drive the tip through a surgical glove into the enclosed hand. Each nematocyst tube may be 10 to 20 times as long as the capsule (i.e. 200 to 300 microns) and carries poison from the capsule into the victim.

Physalia feeds on small pelagic fish. When one is caught the tentacles contract, drawing the prey up within reach of the feeding polyps just under the float.

The potent venom is a neurotoxin which produces paralysis and death in experimental animals by acting first on the nervous system, particularly the respiratory centre and then on the muscles.

Figure 10: Dried blue-bottle tentacles can still sting.

Injury Prevention

- Avoid swimming when blue-bottles are washed up on the beach, as they are likely to be in the water as well. If swimming is unavoidable, protective clothing is a help.
- Do not touch the tentacles of dead blue-bottles on the beach as they remain venomous long after the animal has died, and can cause a painful rash (Figure 10).

Symptoms

- Local blistering and pain, and general paralysis (the purpose of the venom is to paralyse prey rapidly). The effect on humans is determined by the shape of the contact with the tentacle. This is typically one long single line. The strong single tentacle is very adherent, and from the first point of contact it can simply lengthen and wrap around the swimmer. The linear weal has beads where batteries of nematocysts cause particularly strong adhesion. If contact is with the tentacle when it is contracted to a spiral ribbon, the weals can be very irregular in shape – and the intensity of stinging very severe. This can happen when a whole animal is picked up – perhaps by a curious child.

  Royal Australian Navy volunteers demonstrated that if the tentacle or pieces of its tissue are removed from the skin the pain is short-lived, the weal settles in hours and the redness goes within a day, though severe blistering takes longer to heal. However, pain and other symptoms will last for hours if even microscopic pieces of tentacle tissue remain on the skin and nematocysts continue to fire-off.
- A large dose of venom could cause more extensive symptoms – intense pain in the near-by muscles and intense general distress and weakness, particularly breathing difficulty. Such severe symptoms are unusual, and deaths attributed to the Portuguese man-of-war are usually due to something else. The only authenticated death was in the Caribbean, from the much larger, many-tentacled Atlantic species.

  The reports of severe reactions in Australia are due to an allergy, a particular and unusual reaction of a particular person to this species.

Treatment

- This must be directed at preventing further stinging. It must be assumed that there are more unexploded nematocysts adherent to the skin, whether as an obvious tentacle or as microscopic and invisible pieces. The best inactivating agent is vinegar, which should be used liberally. This should make the pain subside rather quickly. If no vinegar is available other methods may be used. (See section on nematocyst stings in general.)

Summary of Treatment for Portuguese Man-of-War Sting

- If tentacle still adherent, pull it off without touching it directly, pulling in the line of the tentacle. It can be wound around a knife or stick like string.
- Douse liberally with 3–10% acetic acid (vinegar), or aluminium sulphate (Stingose).
- In absence of these, protect skin with hot dry sand or oil.
- Rest, elevate and cool.
- In cases of more severe general reaction with severe pain, shock and allergic reactions, medical treatment is necessary.

Fire Coral

Class Hydrozoa
Order Milleporina

Figure 11: The fire coral, *Millepora platyphylla,* which is common on coral reefs in north-western Australia.

Although related to the hydroids and the Portuguese man-of-war, the stinging or fire coral *(Millepora)* looks very different as it has a hard coral-like skeleton. Unlike the true corals however, the surface is very smooth, with minute pores for feeding polyps, surrounded by smaller pores for stinging polyps. The sexual stage occurs in tiny medusae borne in cup-shaped depressions.

In Western Australia the most common fire coral, found on coral reefs off the north-west coast, is *Millepora platyphylla* which forms massive, stony, plate-like colonies, which are yellow-brown in colour with paler edges (Figure 11). There are also several less common bushy species with flattened branches, found in coral lagoons (Figure 12). All have similar stinging properties.

Injury Prevention

- Wear gloves and protective clothing when diving or snorkelling on coral reefs.

Symptoms

- Pain varies from a mild prickling sensation to a severe stinging pain followed by a burning itch with redness, swelling and weals or sometimes blisters. The signs may last more than 24 hours but usually the pain subsides after a few hours.
- Occasionally nausea and vomiting occur.

Figure 12: Millepora sp. in a coral lagoon.

Treatment

- Apply vinegar or aluminium sulphate to the affected area in case unexploded nematocysts are still there.

Since the first contact is usually made with the observer's fingers, keep these well away from sensitive skin, particularly on the face and around the eyes.

Seek medical aid in case of cardiovascular and respiratory problems.

Jellyfish

Classes Cubozoa and Scyphozoa

These jellyfish have an obvious medusa stage, and an inconspicuous polyp stage. They have a central mouth, and most of the medusa bell consists of a thick jelly layer around the stomach. The medusa is saucer, bell or cube shaped.

Most jellyfish drift at the mercy of the currents, often accumulating in large numbers in sheltered bays and estuaries. They swim with a pumping action of their bell but for most kinds this does little more than allow them to move up or down in the water. The four-sided box jellyfish, however, are powerful swimmers with good control of speed and direction.

Jellyfish feed in various ways, depending on the number and length of tentacles and the power of their stinging capsules. Some take in minute planktonic organisms, some are like corals and have algae in their tissues which produces food by photosynthesis, while others catch small crustaceans and fish with powerful stinging capsules on their tentacles.

The stinging jellyfish of north-western Australia are not well known and a number of serious stingings have been reported whose symptoms do not exactly fit those described for known Queensland stingers. Fairly detailed descriptions are given of stinging jellyfish that have been found in north-western Australia in the hope that symptoms from a particular stinging can be related to a definite species of jellyfish. If specimens can be preserved and forwarded to the Museum positive identification can be made. Several small jellyfish which are the medusae of hydroid polyps are described in the section on Hydrozoa (page 11).

Box Jellyfish

Class Cubozoa
Order Cubomedusa

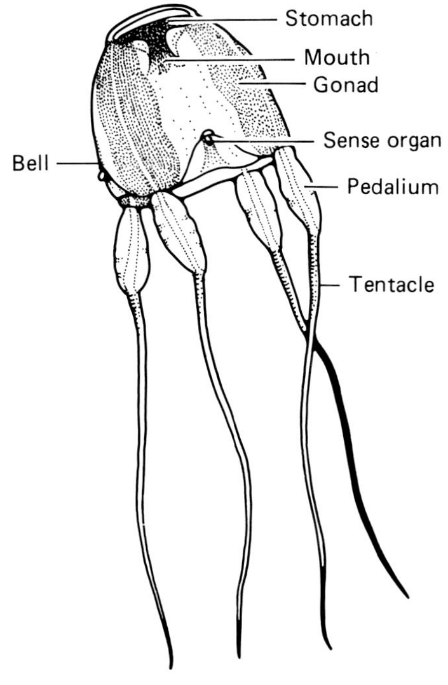

Figure 13: A four-tentacled cubomedusa or box jellyfish which has a squarish bell with each tentacle arising from a leaf-like pedalium.

These have a squarish, box-like bell with one or more tentacles hanging from a tough, gelatinous lobe on each corner of the bell. The lower edge of the bell is folded under to form a shelf. On each side of the bell above the margin is a sense organ consisting of a balance organ and a complex eye, with lens and retina, which looks down into the bell. Inside the bell is a tubular mouth hanging from the centre and gonads which vary in shape and position depending on the species. Box jellyfish are often colourless and difficult to see in the water (Figure 13).

This group includes the most venomous animal in the world, *Chironex fleckeri,* whose severe sting can kill in as little as three minutes. Other species are not deadly but several are severe stingers. Most occur in tropical waters of northern Australia and the Indo-West Pacific, but one species, a relatively minor stinger, occurs in southern Australia as well.

Box jellyfishes are divided into two families, the Carybdeidae, with four tentacles and the Chirodropidae with four clusters of tentacles. Those with many tentacles are the most dangerous.

Several moderately severe stingings by unknown carybdeid jellyfish have been reported off north-western Australia. One at Barrow Island and several at the Rowley Shoals had symptoms similar in some respects to *Carukia* (Irukandji) stings but lacked the usual latent period before the onset of general symptoms. At the Rowley Shoals small, opaque white cubomedusae with bells a few centimetres high and very long tentacles were seen at the time the stings occurred. None was collected so the identity of the culprit remains in doubt.

Small Four Tentacled Box Jellyfish

Family Carybdeidae

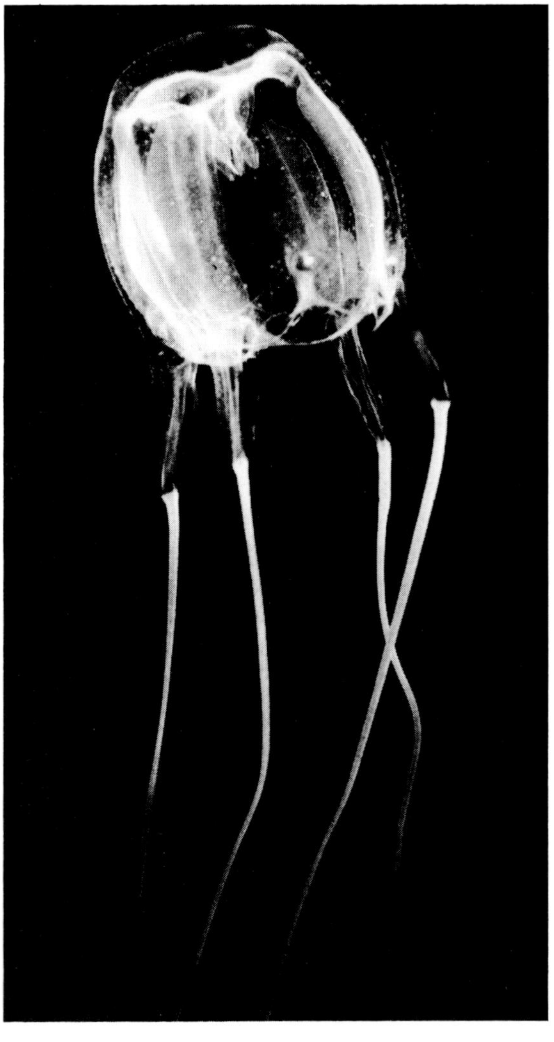

Figure 14: The small box jellyfish, *Carybdea rastoni,* is often troublesome to swimmers in sheltered sandy bays.

Carybdea rastoni

A small box jellyfish, also called stinger, jimble and incorrectly 'blue-bottle' is the second smallest of the box jellies, the bell never growing to more than 3.5 cm long and 2.5 to 3 cm wide, with tentacles 10-15 cm long. The bell is almost transparent so that often the shadows on the sand are more obvious than the animals themselves. The four tentacles may be cream, lavender or pink, and are more easily seen than the bell. The stomach is shallow with a brush-like bundle of gastric cirri in each corner (Figure 14).

C. rastoni frequents quiet bays particularly over white sand where the jellyfish collect close to the bottom when the light is bright in the middle of the day, but disperse through the water or swim near the surface early or late in the day, on cloudy days, and presumably at night.

This small box jellyfish is common in south-western Australia, particuarly in Geographe Bay, off Perth metropolitan beaches, at Rottnest Island and occasionally in the Swan River estuary. It is also found in other parts of Australia and in Japan, the Philippines, Hawaii and other parts of the Pacific. It is most common here in summer and autumn.

Symptoms

- A sharp stinging sensation on sensitive skin may increase or lessen within five minutes, followed by smarting or itching of the area.
- The severity of the sting will vary with the sensitivity of the skin. A sting on the outer side of an arm may be mild but on the armpit or lips or inside a swimsuit it can be very painful.
- Initially up to four red lines from the four tentacles mark the area of contact. The redness spreads and becomes blotchy and a weal 0.5 to 1 cm across may develop. The effects usually disappear in an hour or two though a brown line may remain for several weeks. However, a severe sting may be painful for 12 hours or more with the weal persisting and ulcerating over the next few weeks and with the brown colour lasting many weeks or months.

Treatment

- Refer to page 9.

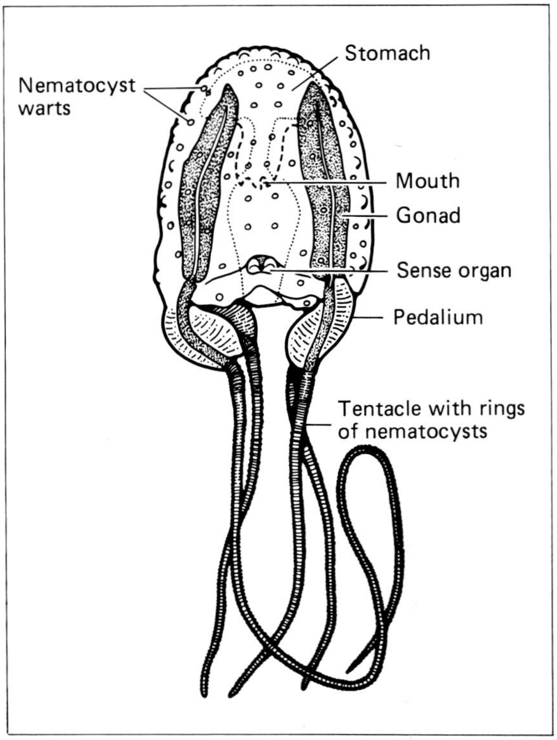

Figure 15: The Irukandji stinger, *Carukia barnesi,* is a thimble-sized box jellyfish with a mild sting which produces severe general effects.

Carukia

Carukia barnesi is known in north Queensland as the Irukandji stinger, after an Aboriginal tribe of the Cairns area. It has a mild sting but causes serious general symptoms (Figure 15).

Carukia has a thimble-sized bell up to 2.5 cm high by 2 cm wide, with four tentacles varying from 4 cm to 1 m long, depending on the degree of contraction. Stinging capsules are grouped in minute red warts on the body and tentacles.

This box jellyfish is found only in tropical Indo-Pacific waters and is responsible for stingings in northern Australia particularly around Cairns, Darwin and the Gulf of Carpentaria. Although it has not positively been recorded from the north of Western Australia it can be expected to occur there as stingings similar to the Irukandji type have been reported from Beadon near Onslow. In the waters around Cairns the Irukandji season begins in late November and continues until about mid-February.

Injury Prevention

- Avoid swimming when jellyfish and gelatinous planktonic organisms are brought ashore with clear oceanic water.
- Wear protective clothing when swimming during the stinger season.

• A few seconds after contact the pain of the sting is sufficient to force the victim to leave the water. Pain increases for a few minutes and diminishes during the next half hour. Within five minutes of contact the area around the sting becomes red and small red lumps appear, increasing for two minutes, before subsiding. Red coloration can last up to three hours and the skin is at first dry then sweats profusely over the area. With the abatement of the initial symptoms the victim might go back into the water and develop further symptoms there. This can be quite dangerous.
• General symptoms are usually felt after about half an hour but it may be between five minutes and two hours. The delayed onset of general symptoms, characteristic of *Carukia* stingings, means that the patient may not relate the later symptoms to the original sting and may be mis-diagnosed as having an acute abdominal illness such as appendicitis or ruptured spleen or even a serious back injury. These symptoms may include.
• Severe abdominal pains with rigidity of the abdominal wall.
• Muscular aches, cramps, joint pains and backache, muscular tenderness and headache which may be severe.
• Respiratory distress and coughing.
• Nausea and vomiting.
• Increased sweating, anxiety and restlessness.
• Still later symptoms include numbness and tingling, itching, smarting eyes, sneezing, joint and nerve pains, weakness, rigors, dry mouth and headache. The temperature usually remains normal. The main symptoms diminish or cease in 4–12 hours with some ill effects persisting for several days. Recovery is complete and no deaths have been reported.
• There is no scarring from the sting.

Treatment

• The intense pain may be relieved with intravenous Pethidine.

Large Four Tentacled Box Jellyfish

Carybdea alata subspecies *grandis*

This species can grow to a large size with a bell 23 cm high and about half that in width. Nematocyst warts may or may not be present on the bell. The four tentacles are each borne on a large lobe or pedalium which has thin flaps either side of a stiff core through which a canal leads to the tentacle. A specimen with the bell 18 cm high has pedalia 5 cm long by 2 cm wide. The stomach is shallow with a crescentic area of short gastric cirri in each of the four corners. The mouth is at the

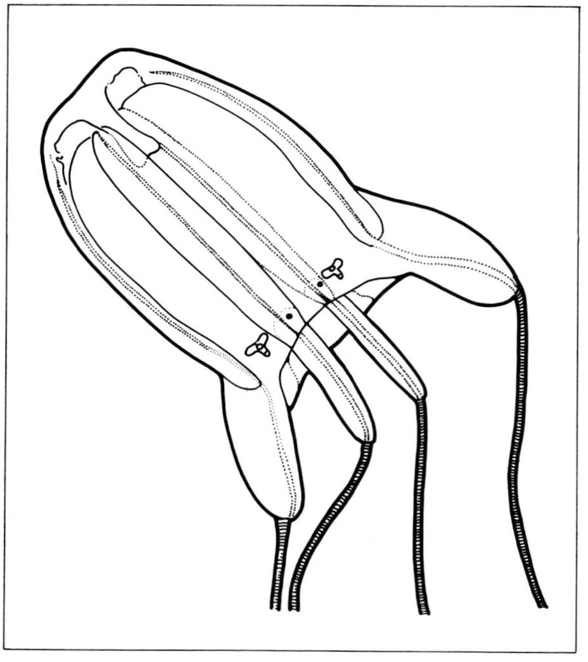

Figure 16: Carybdea alata grandis.

end of a very short tube called the manubrium hanging in the centre of the bell. On each of the four sides of the bell above the lower edge is a dark brown eye spot which is part of a sense organ in a pit enclosed by three flaps, one above and two below (Figure 16).

Several specimens of *Carybdea alata* have been trawled in Western Australian waters north of Port Hedland and one was collected at the surface near Scott Reef, north of Broome. This particular subspecies is widespread in the tropic Indo-Pacific while another occurs in the Atlantic.

Symptoms

• A minor sting, from less than 1 cm of the tentacle of this species, caused intense pain and a weal with surrounding redness. This was followed a few minutes later by tightness in the chest which eased after about half an hour. A major sting from a large *Carybea* could be serious. There was no scarring from the sting.

Treatment

• As for stings from smaller jellyfish (see page 9).

Note: Severe symptoms and tightness in the chest warrant careful watching and medical attention. The use of intravenous Morphine or Pethidine are indicated, and even anti-shock measures summarised as oxygen, intravenous fluids including plasma, intravenous hydrocortisone. Further treatment is in an intensive care situation.

Tamoya sp. cf. *T. gargantua*

Species of *Tamoya* are large, four tentacled box jellyfish with a bell up to 22 cm high by 13.5 cm wide, but they may be nearly as wide as high. A specimen from north-western Australia was 8.5 cm high by 7.5 cm wide. In contrast to *Carybdea* the stomach is deep with four vertical bands of gastric cirri on the walls, and the mouth is at the end of a long tube or manubrium hanging from the centre of the bell.

Another feature which distinguishes *Tamoya* from *Carybdea* is the covering of the sense organs. *Tamoya* has a single flap above and below the sense organ.

Several specimens of *Tamoya* sp. have been trawled from north of Port Hedland off north-western Australia. *T. gargantua* is recorded from the tropical Indo-Pacific including Japan, Malaysia, New Guinea and northern Australia. A man was severely stung on the foot at Exmouth in 1978, and the injury was attributed to a species of *Tamoya* since two specimens (tentatively identified from a photograph as *Tamoya* sp.) were washed up on the beach on the day following the stinging (Endean 1981). However it is possible that the specimens, which had bells up to 10" (25 cm) in height were *Carybdea alata grandis,* also found off north-western Australia.

Symptoms (based on the case above)

- Immediate intense pain followed after about five minutes by a 'knot' in the lower abdomen which progressed towards the chest, restricting breathing.
- Breathing became distressed, with rigid abdominal and chest muscles.
- Intense pain, which seemed to emanate from the base of the spine, spread throughout the body, up the back of the neck to the head.
- Patient became cold and clammy with profuse perspiration.
- After 24 hours the pain was localised in the legs and feet and gradually diminished.
- The patient left hospital two and a half days after being stung but for several days had sore muscles, diarrhoea, bad dreams, chills, twinges of pain in the foot, numbness of skin, weakness, vision problems and headaches.
- Weight loss was about 3 kg. The stung area had 89 small red spots which persisted for nearly four weeks by which time he had recovered apart from weakness in the stung foot and weakened vision.
- There was no scarring from the sting.

Treatment

- The patient treated himself by pouring fresh water over the sting (causing increase in pain through discharge of more nematocysts).
- In hospital he was treated with oxygen then frequent injections of cortisone and painkiller for 24 hours.

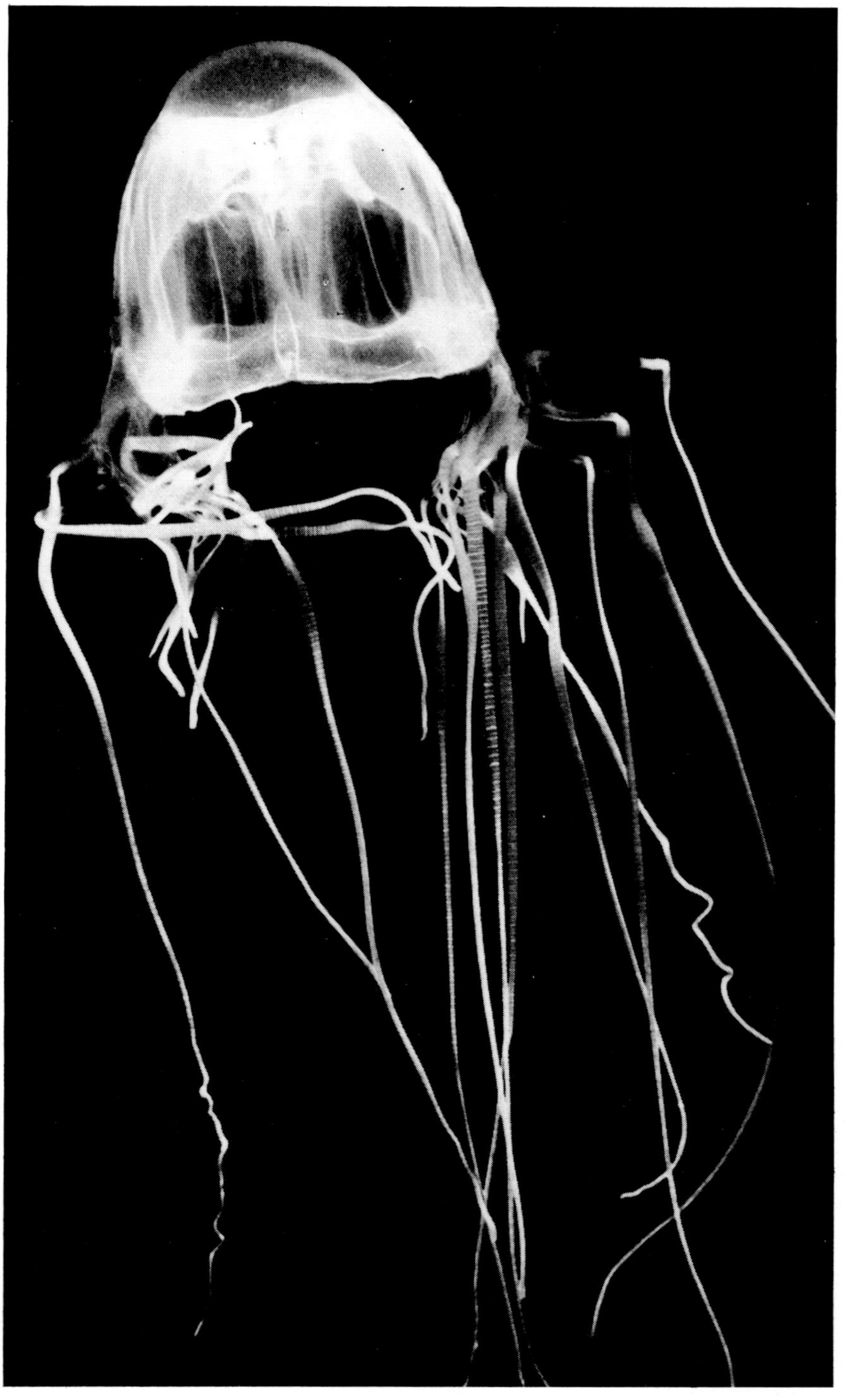

Figure 17: A young specimen of the deadly box jellyfish, *Chironex fleckeri.*

LARGE BOX JELLYFISH WITH MANY TENTACLES

Family Chirodropidae

Figure 18: Chironex is responsible for many deaths in northern Australia. From the sea surface a swimming *Chironex* is almost invisible.

Chironex fleckeri (Figures 17, 18)

'This is the most dangerous jellyfish in the world. Because of it, more than 70 people have met sudden and painful deaths in subtropical Australian waters since 1900. Many of the fatalities were young children who died within minutes of being stung.' (Sutherland 1983). This species was formerly called the 'sea wasp' but this misleading name is no longer used since *Chironex* does not fly and the sting is far more dangerous than that of a wasp.

Chironex is the largest of the box jellyfish, with the squarish round topped bell occasionally reaching the size of a man's head and weighing more than 6 kg. Large specimens are more commonly 10 to 17 cm in bell height with up to 15 tentacles, each up to 2 m long, arising from each of the four claw-like pedalia, while smaller specimens have correspondingly fewer and shorter tentacles.

Chironex was formerly confused with a related box jellyfish *Chiropsalmus quadrigatus*, a less dangerous stinger. The two genera (or groups of species) can be distinguished at any size by the shape of the canals running through the pedalia to the tentacles. In *Chironex* each canal bends at an acute angle projecting upwards, while in *Chiropsalmus* the canal has a right-angled bend (arrows in Figure 19). In immature animals the gonads of *Chironex* resemble cockscombs while those of

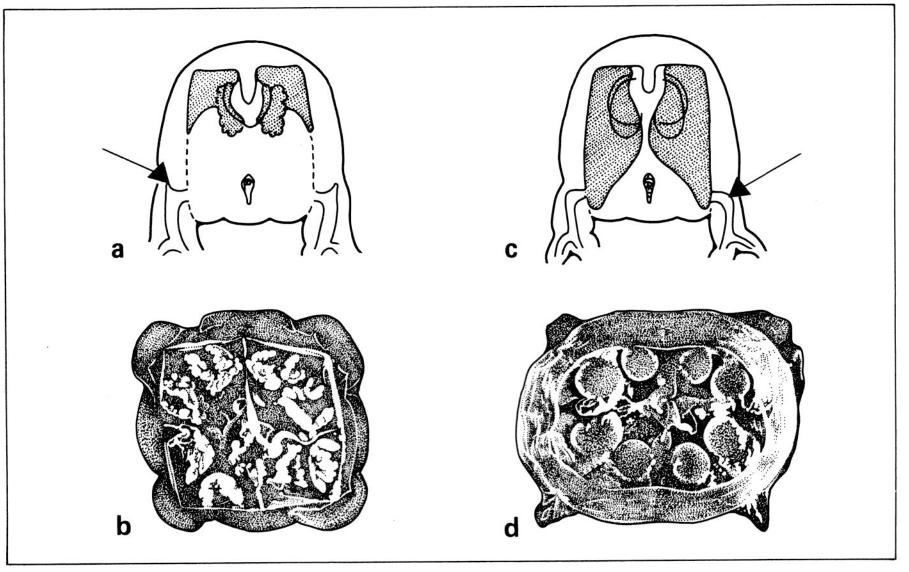

Figure 19: A comparison of the diagnostic features of *Chironex* and *Chiropsalmus* (a) *Chironex* from the side showing the diagnostic acute angle (arrowed) in the canal leading to the pedalium. (b) *Chironex fleckeri,* looking into the bell from below with the tentacles removed. (c) *Chiropsalmus* from the side showing the diagnostic rightangled bend (arrowed) in the canal leading to the pedalium. The gonad is stippled in both. (d) In *Chiropsalmus quadrigatus* the developing gonads are smooth and rounded.

Chiropsalmus are smooth and rounded (Figure 19). When mature the gonads of *Chironex* hang down like bunches of grapes inside the bell.

Chironex and *Chiropsalmus* of the same size have about the same number of tentacles but those of *Chironex* are longer and more heavily armed with nematocysts. A specimen of *Chironex* would have about 10 times the volume of stinging cells of a *Chiropsalmus* of the same size. However *Chironex* grows to a far greater size than *Chiropsalmus* and acquires almost twice the number of tentacles each of which is longer and thicker, so that a well grown *Chironex* could carry 100 to 200 times the amount of venom of a full-sized *Chiropsalmus.*

The largest specimen of *Chironex* reported, the bell of which was slightly larger than a man's head, carried more than 91 metres of tentacles.

Chironex is found in muddy inshore waters of northern and eastern Australia from the Gladstone area in Queensland to north-western Australia. It feeds on a small red shrimp abundant in some areas during the wet season, and on small fish. It seems to favour areas where fresh water dilutes the sea and is found in very shallow water during the wet season in northern Australia. *Chironex* can swim rapidly and change direction abruptly, and so avoid turbulence, remaining seaward of breaking waves. It is thought to go to the bottom in deeper water during rough weather.

Large adults usually appear in November, together with very small juveniles. The juveniles probably mature in three or four months so that the greatest number of mature specimens are likely to be found in March or April. June, July and August are the only months when there are no reports of stingings in Queensland but in the Northern Territory cases have occurred in these months as well.

No fatalities have been linked with this species in the north-west but a child stung at Lagrange Bay, south-west of Broome, in March 1975, was saved by prompt resuscitation. She showed the characteristic symptoms including the heavy scarring of *Chironex* sting, giving fairly positive identification, although the box jellyfish was not collected (Sutherland 1983).

Although *Chironex fleckeri* has not been identified from waters to the north of Australia deaths from jellyfish stings are known from Bougainville Island, the Philippines and Malaysia. *Chironex* is suspected to be the cause of these fatalities.

The Chironex Sting (Figures 20, 21)

Children wading or playing in shallow water, are the most frequent victims. The severity of the sting depends principally on the length of the tentacle in contact relative to the size of the victim, the thickness of the skin in the contact area and the duration of contact. The size of

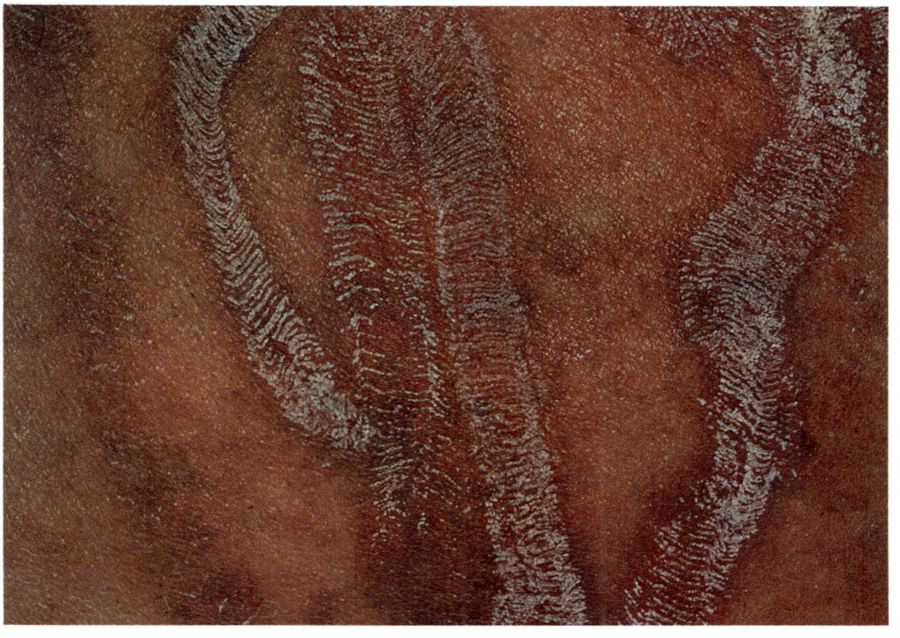

Figure 20: The typical ladder-like pattern of a fresh sting by tentacles of *Chironex fleckeri.* This is followed by skin death of the area and permanent scarring.

the jellyfish and the time since the nematocysts were last used for feeding or defence also affect the amount of venom discharged. Death is probable if the total length of weals on the victim is greater than 6 or 7 metres (Barnes 1966).

The venom contains a protein which is destroyed by heating to 45°C for five minutes and by treating with a protein-splitting enzyme. A traditional treatment in some places is to apply the juice of green pawpaw which contains such an enzyme. The venom has three separate factors causing skin death, destruction of blood cells and a lethal effect on the heart.

Figure 21: Section of skin showing penetration by nematocysts.

Injury Prevention

- In Queensland June, July and August are regarded as safe months for swimming but in the Northern Territory and possibly the Kimberley coast protection should be used at all times in inshore waters.
- Overalls, wetsuits or old clothes give protection. Special 'stinger suits' are available from sports stores in Queensland.
- Avoid even paddling in muddy bays in northern Australia without protection.
- Avoid entering the water if stingings have been reported in the area, regardless of the time of the year.
- The abundance of small red shrimps, *Acetes australis,* in the water or washed up on the beach during the wet season, should be regarded as a danger sign, since these are the preferred food of *Chironex* and *Chiropsalmus.*

Symptoms

- The medical significance of the medusa is profound.
- The severity of the sting of *Chironex* is distinctive and over 70 deaths

have already been documented. Research work has produced the first successful anti-venom against a jellyfish. This does not mean that the problem is solved however, because the speed and severity of the onset of symptoms can cause death before the anti-venom can be injected. The real answer is prevention.

The symptoms relate to the pattern of contact. If the long tentacles of the feeding animal touch a human in shallow calm water they adhere strongly and then contract. Immediately the concentrated batteries of stinging capsules inject large amounts of powerful venom. This is meant to stun small crustaceans. In the human touched by accident, the venom causes great local skin destruction (dermatonecrosis) and spasm of all muscles including those of blood vessels and heart. The pain is thought to be caused by muscle spasm and from tissues denied blood and oxygen because of the muscular contraction of the blood vessels. Death can follow because the heart muscles cannot relax so the heart cannot refill again for the next beat. This produces cardiac shock as in a heart attack. The combination of the heart muscle going into spasm and the arteries going into spasm is thought responsible for the weird fluctuations of blood pressure and general condition of the victim. The stage of heart failure (before it actually stops) produces fluid accumulation in the lungs with difficulty in breathing and even frothing at the mouth.

The clinical state of the victim is very severe. Carefully documented accounts describe him/her in shocking pain, often becoming 'irrational'. The person in danger of death has been in contact with metres of tentacle, i.e., most of the animal. In lethal cases over 16 metres of weals have been measured. The ribbon-like tentacles can be broken off by the animal, (unlike the single tough cord of the Portuguese man-of-war). The victim typically has strands that look like 'apple jelly' or 'uncooked prawn flesh' over the skin, which rapidly develops gross damage resembling irregular, multiple zig-zag whip marks over the stung area.

The difference between life and death depends upon the dosage of venom injected. Cardiac damage tends to be an all-or-none response, which means that below a certain threshold amount of injected venom there is no danger to the heart, but above it there is. The significance is that rescuers may be stung by the strands of jelly they remove from the victim's body but will not be in danger of losing their own lives.

The distraught victim often staggers to the beach, and is likely to collapse. Problems are:

- Keeping him alive.
- Treating the severe pain.
- Treating the skin damage and reducing the risk of future scarring.

A key factor is that the accident is likely to occur far from medical aid, especially in the north-west of Western Australia.

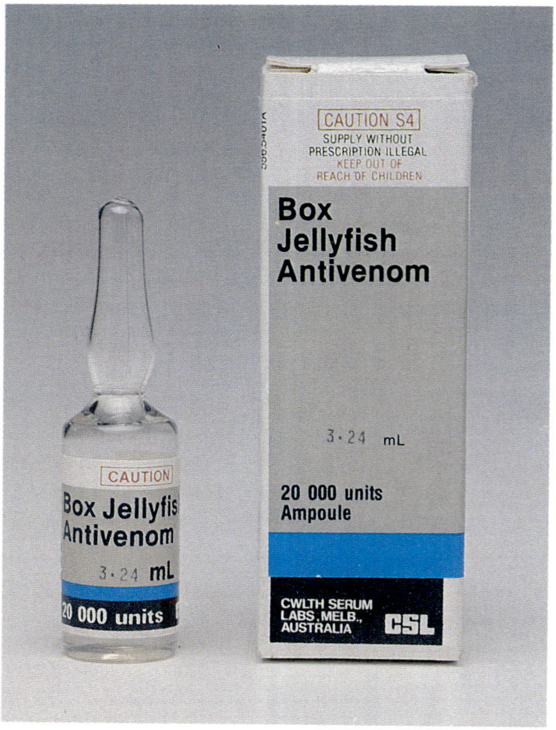

Figure 22: An ampoule of Box Jellyfish Antivenom.

Treatment

- Prevent the continuing injection of venom by inactivating stinging capsules – liberally washing with 3-10% acetic acid (vinegar), literally by the litre. Stingose containing 20% aluminium sulphate could be used, but it is unlikely that sufficient volumes would be obtainable.
- Remove the obvious and highly sticky pieces of tentacle, with a knife, stick or some other tool.
- If respiration is failing keep it going, with any form of artificial resuscitation which works.
- If the heartbeat begins to fade the situation is desperate and external cardiac massage is all that can be done. The hope is that the short-acting venom will wear off, and that if the victim can be kept alive for the first few minutes he will stay alive.
- Get anti-venom to the victim, or the other way around.

In Queensland the anti-venom has been extremely effective, both in pulling people out of a state of intense distress, and for the future healing of the skin.

An ampoule of anti-venom labelled 'Box Jellyfish Antivenom' contains the required dose of 20,000 units in approximately 3 ml and should be diluted 1 in 10 and injected intravenously. Up to 5 ampoules have been found necessary for a single treatment (Figure 22).

- If anti-venom cannot be easily obtained, endeavours to relieve the pain may include:
  - Intravenous Morphine or Pethidine.
  - Long running intravenous drip of Morphine or Pethidine.
  - Regional block anaesthesia.

These require medical facilities with increasing sophistication, the last being done by an anaesthetist.

- In the absence of skilled medical help and equipment.
  - Local anaesthetic cream.
  - Or try Stingose (aluminium sulphate 20%) spread over the marks.
  - Or try papain (the enzyme from pawpaw, either in a fresh green pawpaw, a traditional Philippine remedy, or in steak tenderiser).
  - Oral analgesics.

In the long-term care of the very severe full thickness damage to skin, steroid (cortisone-like) creams may be of value in reducing future scarring. This can be expected to last for months, if not permanently. Queensland reports indicate that anti-venom, even given late, can be a great help in reducing skin damage. However, since it is a foreign protein, which could cause sensitisation, it should not be used for minor *Chironex* stings (Figure 23).

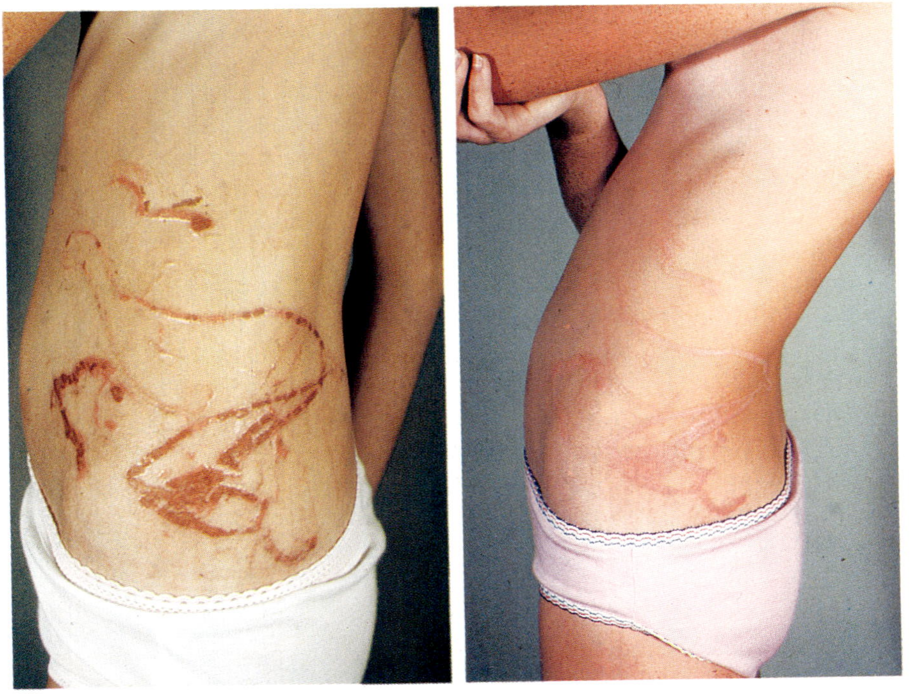

Figure 23: The stung area after one month and seven months.

Chiropsalmus quadrigatus

This species is very similar in appearance and habits to *Chironex* but it is smaller and less dangerous although still a severe stinger. The bell is 7-10 cm high by 8-10 cm wide with a branched pedalium at each corner, each bearing up to nine tentacles which are shorter and thinner than those of *Chironex. Chiropsalmus quadrigatus* differs from *Chironex* in the shape of the pedalial canal which is curved or bent at a right angle in contrast to the acute-angled bend in *Chironex* (Figure 19). There are eight smooth rounded gastric saccules and leaf-shaped gonads with smooth edges quite unlike those in *Chironex* (Figure 19).

Chiropsalmus specimens, tentatively identified as *C. quadrigatus,* have been collected in north-western Australia at Walsh Point in Admiralty Gulf and at Broome, where they have caused the hospitalisation of several people. This species is also found on the Queensland coast and from the Maldive Islands to the Philippines.

The venom has similar effects on the skin, blood and muscles to that of *Chironex* but is less potent having, as detailed before, about 1/100 to 1/200 of the total venom volume of a full grown *Chironex.*

Symptoms

- Stings from up to a metre of tentacle from small specimens are fairly minor and have little general effect.
- A more extensive sting causes severe pain and shock. Several lines of weals surrounded by redness develop. The weals have a pattern of bars corresponding to the rings of stinging capsules on the tentacles.
- Small blisters form along the tentacle markings but there is only shallow tissue destruction and so no permanent scarring even though healing takes a long time. Skin discoloration, at first brownish and later purple usually disappears within six months. The area may itch at times as long as the marks persists.
- General symptoms are similar to those of *Chironex* but less severe and no fatalities have occurred in Australia.

Treatment

- This is as for severe stings – see *Chironex fleckeri* (page 32).

The anti-venom prepared against *Chironex fleckeri* will, in the laboratory, protect mice from the venom of *Chiropsalmus* as well. It is reasonable to hope that it will also give the same protection to humans.

Saucer Jellyfish

Phylum Cnidaria
Class Scyphozoa
Order Semaeostomeae

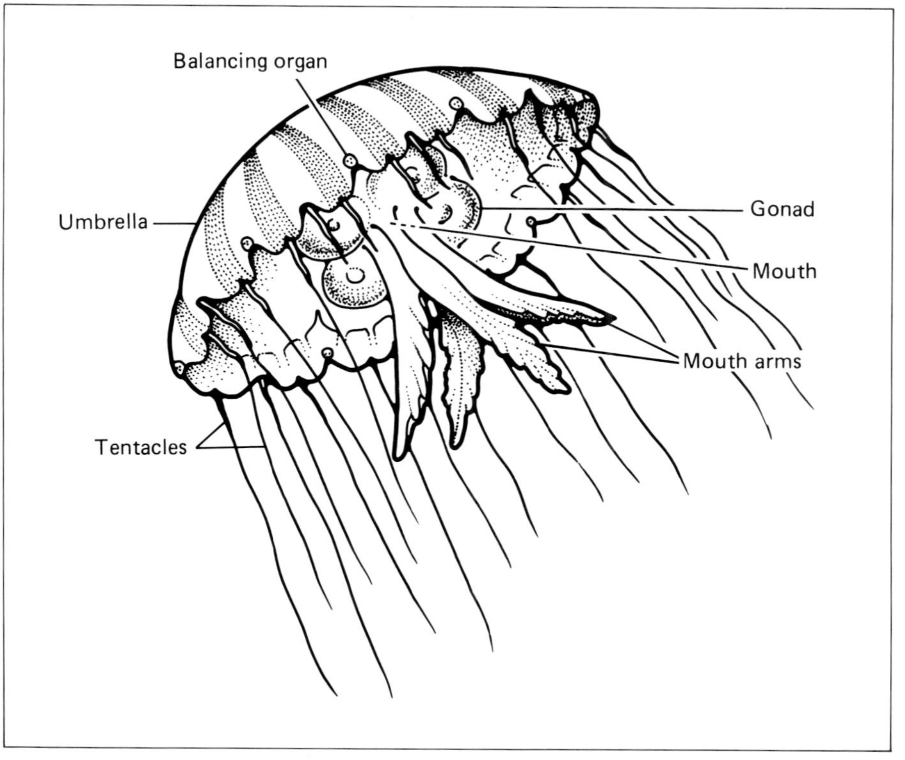

Figure 24: The structure of a stinging saucer jellyfish, *Chrysaora,* shown with some tentacles removed.

Scalloped saucer jellyfish have tentacles and sense organs around the edge and a single central mouth with four large oral arms which fall in long frilly folds. The life cycle includes a minute polyp stage attached to rocks which buds off a succession of tiny jellyfish.

The best known saucer jellyfish is *Aurelia aurita* common worldwide in oceans, estuaries and inshore waters. This species is harmless and can be handled without fear of a sting.

Stinging saucer jellyfish found off south-western Australia, are the mauve stinger *(Pelagia),* the sea nettle *(Chrysaora)* and the hairy stinger *(Cyanea). Chrysaora* and *Cyanea* occasionally enter the Swan River estuary.

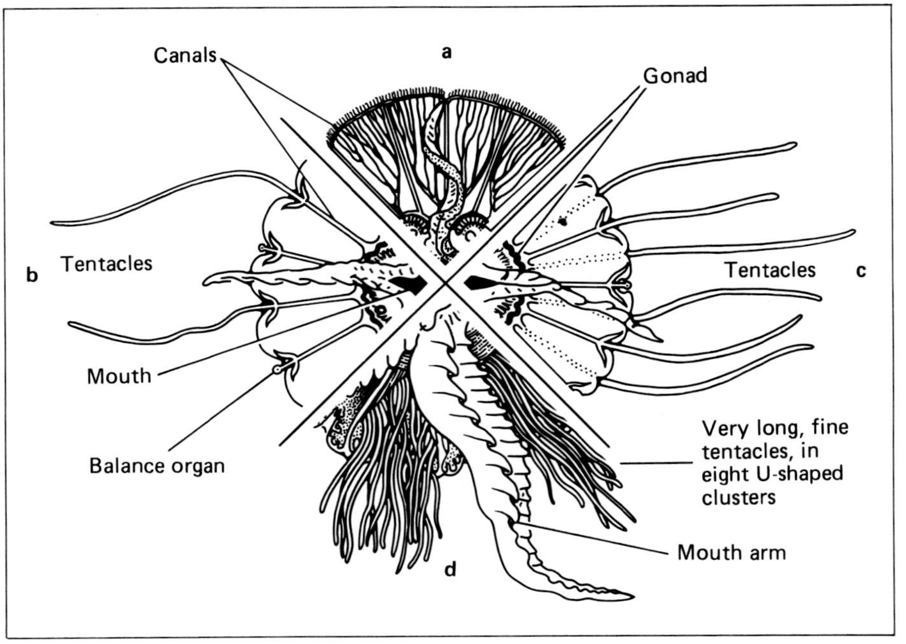

Figure 25: A composite diagram showing the main differences between four genera of saucer jellyfish. In all but *Aurelia* the tentacles have been shortened.

The diagram shows segments of *Aurelia, Pelagia, Chrysaora* and *Cyanea* (Figure 25)

Aurelia (a)

Aurelia has a fringe of very numerous short tentacles around the edge of the umbrella and branching canals leading from the stomach to the margin.

Pelagia (b)

Pelagia has alternating tentacles and sense organs.

Chrysaora (c)

Chrysaora has three tentacles between sense organs.

Pelagia and *Chrysaora* have tentacles arising singly from the margin, simple canals and no ring canal.

Cyanea (d)

Cyanea has very long thin tentacles arising in U-shaped clusters under the umbrella.

Figure 26: The mauve stinger, *Pelagia noctiluca,* has a warty bell, long frilly mouth arms and only eight tentacles.

Mauve Stinger *(Pelagia noctiluca)* (Figure 26)

Tentacles alternate with sense organs. Eight sense organs and eight tentacles. Bell usually about 6.5 cm wide and 3 cm deep.

The umbrella is shallow, up to 10 cm across, with four long mouth-arms hanging from the centre. The umbrella and mouth-arms are spotted with nematocyst warts. There are only eight tentacles, which alternate with the eight marginal sense organs. *Pelagia* varies in colour from brownish-yellow to light rose red, mauve or purple. The colour is strongest on the nematocyst warts, gonads and tentacles. As the specific name meaning night light suggests, it luminesces when stimulated by touching or by wave turbulence.

Pelagia is an oceanic species, found worldwide in warm and temperate waters. It usually occurs in large swarms which are occasionally driven inshore where they may cause mass stingings of swimmers. Large numbers were reported in bays near Cape Naturaliste during 1984 but *Pelagia* can be expected to occur anywhere along the coast. Multiple stingings occur when swarms of *Pelagia* invade a popular beach or a swimming carnival.

Symptoms

- Contact with the outer surface of the umbrella, or with trailing tentacles results in a mild sting although multiple stings may be severely painful. The immediate sharp pain is followed by intense itching which may persist for more than a week.

- In some cases there are small, raised, red lumps which may blister and fester, healing slowly. Extra pigmentation of the skin in the area can last some time.
- Multiple stinging is very occasionally followed by faintness, nausea, vomiting, difficulty in breathing and shock.
- Secondary infection sometimes occurs.

Treatment

- As for moderate stings in which inactivation of unexploded nematocysts and pain relief is indicated (page 9).

Figure 27: The delicate bluish *Chrysaora* sp., often found in Cockburn Sound, is a severe stinger. It has 24 tentacles.

Sea Nettle *(Chrysaora)* (Figure 27)

Three to four tentacles between sense organs. Eight sense organs; 24-32 tentacles. The bell is about 10 cm wide.

This jellyfish is generally larger than *Pelagia* and is often about 10 cm but it may be up to 20 cm across the smooth saucer-shaped umbrella. Long frilly mouth arms hang from the centre of the umbrella which has 24-32 long marginal tentacles (3-4 between adjacent sense organs). The bell is bluish to almost colourless. In Western Australia *Chrysaora* is sometimes abundant in Cockburn Sound and occurs off Perth metropolitan beaches and occasionally in the Swan River estuary. The sea nettle is a severe stinger producing general as well as local symptoms.

Symptoms

● There is a sudden burning sting and then a delay before general symptoms occur.
● Pain spreads from the sting, leading to muscle cramps and excruciating pains across the lower back after about 15 minutes so that patients have been described as being hysterical.
● Restlessness, weakness, and mental depression may follow a severe sting. A characteristic feeling of constriction of the throat and chest with constant coughing and a profuse discharge from the nose and eyes may occur but the symptoms usually abate after about half an hour and disappear after an hour, leaving weariness and sore air passages. In some cases nausea and vomiting can occur but not nearly so much as in the Irukandji sting which is similar in its delayed onset.
● The sting is followed by grossly raised red weals, which swell, weep like eczema and blister, taking two or three weeks to heal, in contrast to the spreading red area but minor skin damage of the Irukandji sting.

Treatment

● Removal of the tentacle, which is often visible, and inactivation of unexploded nematocysts is indicated, as is the relief of pain as in all severe stings.

Hairy Stinger or Snotty *(Cyanea)* (Figure 28)

The name 'Snotty' comes from the mucus which is copiously secreted by the mouth arms and which has a fishy odour. The upper surface of *Cyanea* resembles the harmless common saucer jelly, but there the resemblance ends. Very numerous long, thin tentacles are grouped in eight U-shaped areas on the underside. Surrounding the central mouth are the greatly-divided, frilly mouth arms which give rise to one description of this jellyfish as resembling 'a mop hiding under a dinner plate'.

In Australia *Cyanea capillata* grows to 1 m wide, whereas it can be twice as large in cold European and sub-Antarctic waters. Most *Cyanea*

Figure 28: The north-western Australian hairy stinger, *Cyanea mjobergi,* photographed off Dampier.

are 15-20 cm across the umbrella with more than 1,000 tentacles, each capable of stretching to 3 m long. Large specimens are occasionally washed up on beaches on the south coast of Western Australia. Large *Cyanea* have been sighted in Queensland waters by the pilots of light aircraft. Small hairy stingers, about 10 cm across are sometimes found in the Swan River estuary. These have a slightly yellowish umbrella with yellow brown mouth arms and probably belong to a Western Australian species *C. mjobergi,* first found off Cape Jaubert in north-western Australia. This species only grows to about 14 cm across and is moderately common in the north-west.

Cyanea capillata is found worldwide in temperate and cold seas while *C. mjobergi* is only known from Western Australia.

Small *Cyanea* (10 cm wide bell) sting only mildly or may even fail to sting when touched, but their stinging ability varies. Large ones sting severely and a massive sting from a large specimen can be dangerous. Detached pieces of tentacle or tentacles trailing metres away from the body of the jellyfish are almost invisible and hard to avoid. Even dried tentacles on fishing nets are capable of stinging when re-moistened.

Symptoms

- The sting produces a sticky feeling followed by burning and then severe pain, which varies with the amount and length of time of contact. The pain may last an hour or more but the swelling fades rapidly though the red marks can last for days. Where there have been knots or crossing on tentacles there can be more severe damage to the skin, with blistering which heals slowly.
- A straight-line sting can occur from a piece of detached tentacle but more often, because of the large number of fine tentacles, the weals are multiple but narrow, often in parallel lines or zigzags, since the tentacle may attach to the victim's skin at the point of initial contact and then contract across the skin. A tangle of tentacles might cause a bizarre pattern of weals.
- Respiration and heart action may be affected by a severe sting. There can be great difficulty breathing, the patient may note that his heart is racing or apparently missing a beat, and he may feel severe chest pain.
- Profuse sweating, mucus secretion from the nose and throat, and muscle spasms have been described. (In the 19th century these symptoms were deliberately induced in Norway as part of jellyfish 'bath therapy'!)
- Eye injuries from contact with *Cyanea* tentacles are sometimes caused by persons swimming underwater with their eyes open. Pain is immediate and severe, the eyelids swell and superficial or deeper lesions to the cornea can occur. Though, with treatment, these usually heal within a week without any residual effect, such eye injuries should be seen by an eye specialist or by a doctor experienced in treating such injuries and using opthalmic magnifying equipment.

Treatment

- As for jellyfish of lower severity (see page 9).

A toxin has been isolated from *Cyanea* which affects the heart of experimental animals. Some fishermen are reported as having developed an immunity to *Cyanea* stings, though this is a rare event with marine venoms.

Rhizostome Jellyfish

Order Rhizostomeae

Rhizostome jellyfish have a scalloped edge without marginal tentacles and a denser, more solid umbrella jelly than saucer jellyfish. There are eight mouth arms, greatly subdivided, with numerous tiny mouths instead of a single one at the centre. Tentacles hang from the mouth arms in some species. None of the rhizostomes are severe stingers.

The commonest rhizostome jellyfish in Western Australia is the spotted jellyfish, *Phyllorhiza punctata* (Figure 29). Several as yet unidentified species of rhizostome jellyfish occur in the north-west, sometimes in large numbers.

Figure 29: Phyllorhiza punctata, a mild stinger, is abundant in southern Australian estuaries. The finely divided mouth arms have tentacle-like filaments but no tentacles hang from the edge of the umbrella.

Spotted Jellyfish *(Phyllorhiza punctata)*

This brown jellyfish which has white spots on its umbrella occurs in large numbers in sheltered bays and estuaries in south-western Australia and is particularly abundant in the Swan River estuary.

The smooth yellow brown umbrella of dense jelly may be up to 50 cm across, although it is usually much smaller. It has eight sense organs but no tentacles around the edge, and the eight branching mouth arms, hanging from the centre, end in tentacle-like filaments. The brown colour is due to single-celled algae which live in the jellyfish tissues.

The fertilised eggs of this jellyfish adhere to its mouth tentacles where they grow into the polyp stage which buds off minute jellyfish.

Phyllorhiza punctata is known from eastern and western Australia, the Philippines and Japan.

Symptoms

- The spotted jellyfish can often be handled without giving a sting but it is capable of mildly stinging sensitive parts of the body. In one case, stinging on the eyes caused blindness for two and a half days. In another case a brown jellyfish stung an eye which became sore, the conjuctiva turned red, and multiple barbed stings were seen over the cornea. Fifty stings were removed in one operation, and it took over a month and two operations before the inflammation of the cornea subsided.

Treatment

- (see page 9).

Figure 30: Pseudorhiza haeckeli is a mild stinger, more often found offshore than in estuaries. The reddish net-like pattern on the umbrella and the single long palp distinguish this species.

Net Patterned Jellyfish *(Pseudorhiza haeckeli)* (Figure 30)

Pseudorhiza is not seen very often but its striking coloration attracts attention. The umbrella is covered by soft nodules outlined by a fine copper-red network. It can grow to about 25 cm across though most specimens are 10-20 cm across. The mouth arms are finely divided but more delicate than those of *Phyllorhiza* and a single long, three-sided palp hangs from one of the eight mouth arms. The palp is a brilliant blue tipped with salmon pink.

Pseudorhiza is found in southern Australian waters from Western Australia to Victoria. Though usually found offshore it does occasionally enter the Swan River estuary.

Symptoms

- Under certain conditions, it may give a mild sting, persisting for half an hour, but it is usually regarded as a non-stinger.

Treatment

- (see page 9).

Sea Anemones and Corals

Class Anthozoa

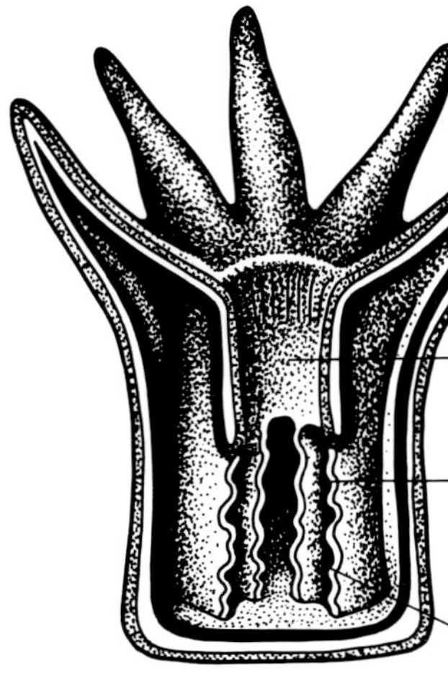

Figure 31: A sea anemone cut away to show the body cavity which combines the functions of digestive and circulatory systems and carries waste and reproductive products out of the mouth, the only opening.

These 'flowers of the sea' are often brightly coloured and attractive but they should be treated cautiously as they are related to jellyfish and do have stinging capsules. The sticky feel of anemones is due to their stinging cells adhering to the skin, and a few species are severe stingers.

A sea anemone (Figure 31) has a tubular body with a single cavity that combines the function of digestive and circulatory systems and also carries waste and reproductive products which pass out of the mouth, which is the body's only opening. Internal partitions support muscles and gonads and increase the area for food absorption. Nematocysts are borne externally on the tentacles and internally on special threads attached to the partitions.

Although anemones often appear fixed to the rocks, most have a muscular basal disc with which they can move about.

Another group of anthozoans called zoanthids, related to anemones, do not sting but are among the most poisonous animals in the sea.

Stony corals are virtually anemones which make a skeleton. They rarely sting but often are the cause of cuts which are slow to heal and may cause coral poisoning.

Injury Prevention

- Treat all anemones with caution.
- Wear gloves if you handle them.

Treatment

- Apply vinegar then remove any adhering tentacles.
- Apply a local anaesthetic ointment such as Lignocaine.
- Obtain medical attention for severe stinging.

Anemones

Subclass Ceriantipatharia

Figure 32: Cerianthus, its column encased in a tough, fibrous tube, lives in sand or mud.

The *Cerianthus* Anemone (Figure 32)

This anemone is more closely related to the black corals than to true anemones. The long soft body, pointed at the bottom, is enclosed in a tough, fibrous tube about 25 cm long by 3-4 cm wide, embedded in muddy sand or mud. When expanded these animals have a double crown of long slender mauve or white tentacles, and in large groups they resemble a field of underwater flowers.

Cerianthus do occur intertidally in mud flats in north-western Australia but are usually subtidal. One species is common in Cockburn Sound, at a depth of 5-20 metres.

A *Cerianthus* sp. in New Caledonia is described as having stinging tentacles which 'inflict excruciating pain' (Catala 1964). The stinging powers of Western Australian species are not known.

Figure 33: The large Cockburn Sound anemone, *Dofleina armata,* has prominent beads on the tentacles containing powerful nematocysts.

Subclass Zoantharia
Order Actiniaria

Cockburn Sound Anemone *(Dofleina armata)* (Figure 33)

The squat tube-like body of *Dofleina armata,* about 6 cm high by 4 cm wide, partly buries in muddy sand or mud at depths of from one to 20 metres. The top of the body has a central slit-like mouth and thick, tapering tentacles around the edge, studded with oval beads heavily armed with potent stinging capsules. The tentacles have a spread of 15 cm or more and wave actively, searching for prey as large as a 10 cm long blow-fish *(Tetraodon).*

The body is light fawn with fine, dark brown vertical lines; the tentacles are rich brown above and below and with white sides, or fawn with irregular bands of brown and white nematocyst beads.

Originally described from Japan and known from the Philippines, *Dofleina armata* has been found in Western Australia only in Cockburn Sound, in the estuary near Fremantle and Shark Bay.

Symptoms

- Intense burning pain and weals patterned with white dead skin in the shape and pattern of the batteries of nematocysts.
- Pain similar to that from an open burn persists for several hours and soreness remains for 36 hours or more.
- Small blisters appear after five hours, with redness and swelling around the stung area. A sting on the back of the hand produced scabs after 48 hours and scars which were still visible after two months.
- If the skin is sensitive in the stung area ulcers may occur, taking several weeks to heal. The ulcers may become secondarily infected.
- General symptoms may be fever, nausea and vomiting, headache and, in very serious cases, shock.

Treatment

- Inactivate any residual nematocysts as usual, and attempt pain relief. This may necessitate the use of long-acting anaesthetic. Secondary infection can be treated with antiseptics and antibiotics.

Stinging Anemone *(Actinodendron plumosum)*

This large grey/green anemone, with branching tentacles, lives subtidally in sandy areas adjacent to coral reefs. It is able to withdraw completely into the sand leaving a hole about 4 cm wide. It has been found in north-western Australia southwards to the Abrolhos but it is uncommon. In Queensland this species is reported to cause very painful stings remaining sore and swollen for more than a week. The red rash of small blisters was visible for a month. (Cleland and Southcott 1965).

Treatment

- As for Cockburn Sound anemone (above and page 46).

Figure 34: A giant anemone, *Stichodactyla* sp., has an undulating disc about 30 cm across covered with very short, strongly adhesive tentacles.

Giant Anemones *(Stichodactyla* spp.*)* (Figure 34)

These giant anemones have an undulating oral disc 30-50 cm across, covered in very short tentacles. There are several species on coral reefs in the Indo-West Pacific, and three occur in north-western Australia southwards to the Abrolhos. These species were formerly included in the genus *Stoichactis.*

The tentacles of *Stichodactyla haddoni* are sticky to touch and may fasten to human skin so strongly that they pull away from the anemone. Contact with them is painless but can raise welts. *S. gigantea* has even more adhesive tentacles but the nematocysts cannot penetrate human skin.

At least two species of *Stichodactyla* are known to produce venom that is hemolytic, so any giant anemones should be treated cautiously.

Zoanthids

Order Zoanthidea

Zoanthids are anemone-like polyps a few centimetres high usually joined at the base or completely fused to one another to form colonies. They do not sting but species of *Palythoa* are highly poisonous.

Figure 35: Palythoa heideri, a highly toxic zoanthid. In Hawaii a related species was traditionally used to poison spear tips.

Palythoa spp. (Figure 35)

Sand grains embedded in the outer wall give *Palythoa* a sandy colour and texture. The polyps of *P. heideri* are about pencil thickness, a few centimetres long, and when expanded are topped with green or brown oral discs fringed with very short tentacles. In *P. densa* the polyps are fused into an encrusting mat-like colony dotted with polyp openings.

This worldwide tropical genus has several species in north-western Australia while two, *Palythoa heideri* and *P. densa,* occur on limestone platforms particularly at Rottnest Island and near Margaret River.

The toxicity of Australian species has not been investigated but a highly poisonous substance, palytoxin, has been isolated from species of *Palythoa* in the Pacific islands. It has the greatest known toxicity of any marine toxin, many times more toxic than puffer-fish poison.

In Hawaii, one species was traditionally used to poison spear tips. The toxin accumulates in the ovary and is lost when the eggs are released. These eggs cause temporary toxicity in bivalves and fish which eat them. A closely related toxin occurs in a Caribbean species and the Australian species are also expected to be toxic.

The poison acts particularly on the heart and circulatory system when tested by injection and causes lesions when administered by mouth.

Injury Prevention

- *Palythoa* should not be handled. Take care that a cut surface of *Palythoa* does not touch any cuts or scratches.

Stony Corals, True Corals

Order Scleractinia

Goniopora (Figure 36) and *Plerogyra* (Figure 37) are among the few stony corals reported to sting.

The greatest danger is from coral cuts which are often hard to heal and may produce an allergic reaction. Coral stings should be treated as for jellyfish stings (page 9).

Figure 36: The long polyps of *Goniopora,* expanded during the day, completely hide the stony skeleton. Species of *Goniopora,* common on Western Australian coral reefs, have a mild sting and produce a toxin.

Figure 37: Plerogyra sinuosa, found on coral reefs off north-western Australia, is readily recognised by its inflated bubble-like vesicles, expanded during the day. At night these are withdrawn and tentacles are expanded. *Plerogyra* has a moderate sting.

Figure 38: Plates of *Acropora* coral break easily and may lacerate the unwary diver or wader on a coral reef.

Coral Cuts

The greatest danger from corals is that of being cut by the sharp edges of their skeletons, especially if these are broken (Figure 38).

When people are walking on reefs the coral often collapses underfoot lacerating ankles and legs. Skindivers may be thrown against coral by waves but even brushing against corals can cause scratches that are hard to heal. This is due to the fragments of coral tissue and mucus from the coral which often remains in a cut. The body reacts against this foreign matter so that subsequent coral cuts or scratches can produce a severe allergic reaction.

Injury Prevention

- Wear boots when reefwalking, and gloves when handling corals.
- Be careful when stepping out of boats on to reefs and avoid stepping on plate corals which are deceptively fragile.
- Wear protective clothing and flippers with heel protection when diving on coral reefs.
- If you become increasingly reactive to coral cuts, it may be necessary to keep away from coral.

Symptoms

- A small and often clean looking cut may begin to sting a few hours later, especially during washing. There may be mild inflammation around the cut.
- Within the next day or two the inflammation may spread, with local swelling, redness and tenderness on light pressure and movement.
- The cut may develop into a festering sore.
- Lymph glands in the groin or armpit may swell and become tender and there may be pain or aching in adjacent joints. If properly treated these symptoms will disappear after a few days but may continue for months if left untreated.
- Itching may persist for weeks or months.

Treatment

- Thoroughly clean even the simplest and apparently cleanest wound as soon as possible after injury with antiseptic lotion (hibitane, Cetavlon, iodine, soap, Phisohex etc.) and remove all foreign material, even using a brush if necessary.
- Apply antiseptic, antibiotic powder or ointment (e.g. neomycin) as soon as possible. If the person is planning to stay near the water, Plastic skin is useful to protect the cut but dry dressings are best if the victim can stay away from the water.
- If the inflammation spreads, obtain medical attention.

Bacterial Infections of Marine Injuries

There are other mechanisms which can cause maritime injuries to heal slowly. A low grade infection can occur because of other life forms on the main creature, such as the small anemone which gives sponge divers' disease; the bacillus *Mycobacterium marinum* which can cause tremendous lumpy swelling in and under the skin; and the bacterium *Erysipelothrix* which causes crayfish handlers' disease. Tetanus and gas gangrene are also possible serious complications of marine injuries.

Mycobacterium marinum

Is related to the organism which causes tuberculosis, and, like it, is very slow growing and produces a granuloma or large lumpy mass. It can cause infection in and under the skin. The easiest to diagnose are large lumpy non-healing ulcers. More difficult ones are lumps in limbs and around joints, often occurring a long time after the initial penetrating injury. They have been confused with forms of arthritis and ordinary bacterial infection. Ordinary antibiotics and anti-arthritic drugs fail to work.

Treatment

- Surgical removal of the mass may be necessary, and anti-tuberculous drugs.

Tetanus *(Clostridium tetani)*

The disease comes on as muscle pain, weakness and spasms coming on days after the injury and requiring intensive care. It is very serious.

It is linked with penetrating wounds, foreign matter and destroyed flesh (of the victim).

Treatment

- Surgical 'debridement' – thorough cleaning of foreign and dead matter – and booster vaccination or anti-serum and Penicillin.

Gas Gangrene *(Clostridium welchii)*

This putrefaction of dead muscle tissue causes grave illness. Gas gangrene requires intense treatment based on the same principles as those applying to Tetanus – debridement, specific anti-serum and Penicillin.

Fish Handler's Disease *(Erysipelothrix insidiosa)*

Coral scratches may lead to a bacterial infection, also known as crayfish poisoning or coral poisoning. The organism is found worldwide, and causes discomfort and sometimes illness among fish, whale and meat handlers. *Erysipelothrix* is a bacterium which enters the body through breaks in the skin. Punctures from rock-lobster and fish spines are particularly prone to infection.

Injury Prevention

- Wear tough gloves when handling rock-lobsters or fish.
- Wear boots when wading or fishing.

Symptoms

- After injury to the skin, there is little reaction for one to seven days, during which the injury apparently heals.
- A purplish red circular area then develops around the puncture, spreading outwards at about 1 cm per day. Itching, burning or pain is felt, followed by swelling. More severe symptoms include fever and joint pains.
- If the area is protected against injury, but otherwise untreated, it will heal within three weeks. However if it is further damaged, secondary infection may result in the development of a boil or carbuncle.
- Rare severe complications include septicaemia (blood poisoning) and the severe version of this, endocarditis (involvement of valves of the heart) which is a risk to life.

Treatment

- Treat all small marine cuts, particularly puncture wounds, seriously.
- Apply antiseptic solution immediately.
- Apply local antibiotic (e.g. Neomycin) powder or ointment four times a day.
- Keep the area clean and dry.
- Avoid further damage to the area.
- A course of Penicillin should be prescribed.

Flatworms

Phylum Platyhelminthes

The Platyhelminthes or flatworms have a flat, unsegmented body, with a distinct head and tail end, and a mouth but no anus. The Turbellaria are free-living, some in damp forests, others are brightly coloured marine flatworms which, incidentally, produce a defensive toxin.

Most Platyhelminthes are parasitic such as the tapeworms (Cestoda) and the blood and liver flukes (Trematoda). Several species of blood flukes *(Schistosoma* or *Bilharzia)* cause the serious human tropical disease schistosomiasis. The parasite passes through an aquatic snail host, multiplying on the way, before infecting humans but fortunately it is not present in Australia.

Other schistosomes infect birds, after passing through a snail first, and these may accidentally infect humans although they cannot develop further and die in the skin, setting up a local irritation.

Schistosome Dermatitis or Swimmers' Itch

Class Trematoda

Since the early 1960s people have been infected by swimmers itch in the Swan River estuary. It has been traced to a parasitic flatworm or blood fluke, *Austrobilharzia terrigalensis,* whose adult stage lives principally in silver gulls while the larval stages live in an estuarine snail called a creeper *(Velacumantus australis)* (Figures 39 and 40).

Figure 39: The secondary host of *Austrobilharzia terrigalensis* is the estuarine snail *Velacumantus australis*. Common on sand flats in the Swan River estuary.

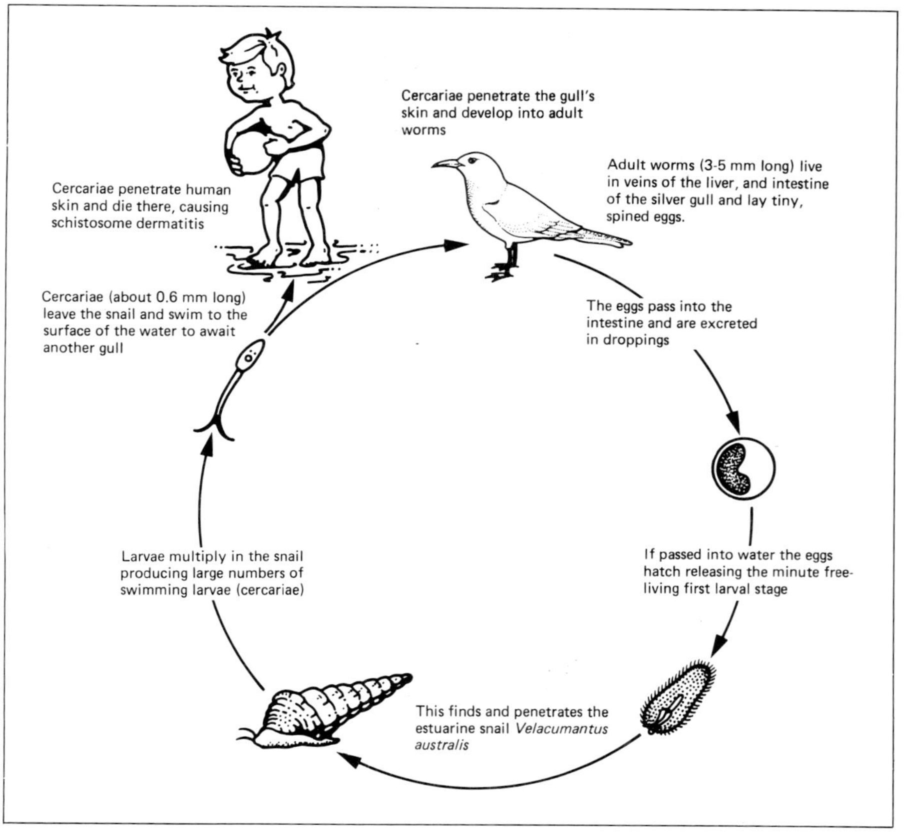

Figure 40: The life cycle of the parasite, *Austrobilharzia terrigalensis,* which causes swimmer's itch in the Swan River estuary.

Eggs of the parasite are excreted in the bird's droppings and hatch into swimming larvae which burrow into the soft body of the snail where they multiply through several larval stages. The last larval stage, known as a cercaria, emerges from the snail and swims to the water surface where it can live for a day or two. If it comes in contact with a silver gull it penetrates the skin and migrates in the blood to veins in the liver, where it matures and mates, finally settling in veins near the intestine. Its spiny eggs pass through the wall of the intestine and are passed out with the bird's faeces to start the cycle again.

Dermatitis occurs when cercariae penetrate the skin of sensitive persons, particularly children. The cercariae cannot develop further in humans but die in the skin, setting up an irritation there (Figure 41).

The river beaches most popular with swimmers, particularly families with young children are, unfortunately, also the ones most used by both gulls and the snail. These creepers are abundant on shallow weed covered sand flats in the Swan River estuary between East Fremantle

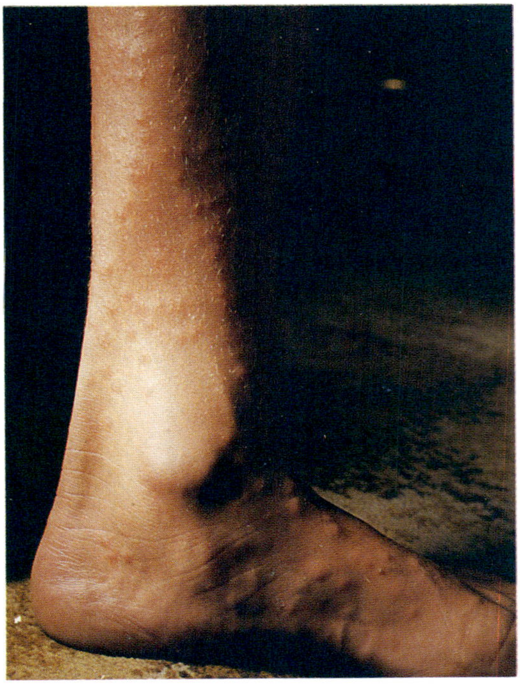

Figure 41: Schistosome dermatitis causes intensely itchy red lumps on any part of the body that has been immersed in infected water.

and Heirisson Island and in the Canning River as far upstream as Mount Henry. The snail is an immigrant, believed to have arrived in the early 1950s from eastern Australia, where swimmers' itch has long been suffered by swimmers in coastal lagoons. While it is abundant in the Swan estuary it is known elsewhere in Western Australia only from Woodman's Point. A few cases of dermatitis have occurred at Mandurah but *Austrobilhartzia* has not been definitely implicated. A type of swimmers' itch is also caused by cercariae of other species of flukes in freshwater lakes in the wheatbelt, and in lakes and rice fields in the Kimberley region.

The risk of swimmers' itch is highest during summer months, particularly when the water over shallow sand flats is warmest. Larvae are released from snails mainly during early morning and late afternoon although most infection occurs in the late morning, and people prawning at night are also sometimes affected. The snails and larvae occur close to shore in water less than a metre deep so there is less chance of infection when swimming from a jetty than when wading from the shore. Larvae are not released from snails for a few weeks in mid winter in the lower estuary, and when the salinity is lowered by winter rain in the middle estuary, usually between June and October. However only about 2% of snails are infected and a spot survey showed only about 7% of people in contact with the river were affected by swimmers' itch during the summer.

Injury Prevention

- Avoid swimming or wading on river flats where the host snail is abundant and where seagulls concentrate.
- Brisk rubbing with a towel immediately after leaving the water may remove some of the cercariae which have attached to the skin.
- If you have to spend much time in infected waters rub the skin with copper oleate (12.5% in yellow soft paraffin). Dimethylphthallate solution (25% in anhydrous lanolin) may be used but this damages plastics and vehicle upholstery.
- Wear boots, long trousers and a long sleeved shirt when wading after crabs or prawns.

Symptoms

- Half to one hour after contact a small red spot, about 1 mm across appears where each cercaria has entered the skin.
- These become intensely itchy red lumps up to 5 mm in diameter usually within 12-24 hours. The dermatitis only appears on parts of the body that have been immersed in infected water so the distribution of spots in relation to clothing worn when wading or swimming usually gives the diagnosis, together with the story of water contact and then the long-lasting nature of the patchy rash. However if the pattern is not obvious there can be confusion with chicken pox or impetigo (school sores).
- The patient may have only a few spots or as many as 200.
- In severe cases the lesions look like eczema, or may form blisters, and there may be swelling of the ankles.
- Severe itching lasts for two to four days and the dermatitis for one to two weeks, or much longer if secondary infection occurs.
- Discoloration of the skin often persists for several weeks after the lumps disappear.

Treatment

- Calamine-based creams, some of which contain a local anaesthetic, give some relief from the severe itching. These may be bought without prescription.
- Anti-inflammatory and anti-allergic steroid ointments or anti-histamine preparations may be prescribed by a doctor.
- If the lesions are scratched and become secondarily infected, broad spectrum antibiotics may be required.

Swimmers' Itch from Algae

Swimmer's itch can also be caused by a blue-green alga, *Microcoleus lyngbyaceus* which grows as a mat of fine brown or dark bluish-green threads on sand, or attached to sea grasses or algae. The alga is found worldwide but toxic effects have only been reported from the Ryukyu Islands and Hawaii. It may affect swimmers in Western Australia but has not so far been implicated.

Injury Prevention

- Avoid swimming where there are drift lines of detached algae and scum. Take a shower, preferably hot, as soon as possible after swimming as the toxin is inactivated by heat.

Symptoms

- Within several mintues to a few hours after swimming people who remain in bathing suits, without taking a shower, may suffer itching and burning sensations, followed by rashes, small blisters and finally peeling skin after four to six days.

Treatment

- Take a hot shower.
- Apply calamine lotion to the affected area.

Bristle Worms

Phylum Annelida
Class Polychaeta

Figure 42: The spotted bristle worm, *Chloeia flava,* has abundant fine bristles, irritating to the skin.

Annelid worms include earthworms, leeches and marine bristle worms (Polychaeta). They have elongated, ringed (segmented) bodies with a mouth at one end and anus at the other. They all have well-developed digestive circulatory and nervous systems and the Polychaetes have bristly projections from each segment which serve as feet, paddles and gills. The bristles are sharp and have a protective function, as well as helping the animal to move. Most Polychaetes are harmless but particularly bristly ones can cause problems and some give a nasty bite.

Worms with Venomous Bristles

Chloeia flava (Figure 42)

This beautifully spotted oval species up to 7 cm long, lives on silty sand bottoms in the tropical Indo-West Pacific and northern Australia. In Western Australia, it is found as far south as Shark Bay, where it is moderately common. It is attracted to baited fishing lines and is sometimes brought up in this way. It should be removed from the hook with gloves as it has many fine bristles which can embed in the fingers.

Figure 43: The salmon-pink fire worm, *Eurythoe complanata,* has hollow bristles containing poison which causes blistering and numbness if touched.

Eurythoe complanata (Figure 43)

This salmon-pink fireworm, with a thick fringe of white bristles on either side, can be up to 14 cm long and 1-2 cm wide (including bristles) but is usually much smaller. This circum-tropical species is common on coral or limestone reefs, under boulders and in the weed mat. In Western Australia it is found on the north-west and west coast and southwards to Cape Naturaliste.

The hollow calcareous bristles contain venom.

Sea Mouse *(Aphrodita australis)*

This 15 cm long, bristly mouse-shaped work is sometimes washed up on beaches in southern Australia and is picked up because of its peculiar appearance. Fine bristles form a felt-like covering over its back while stiff bristles project from its parapodia or feet.

It has been reported as causing a 'nettle sting', from the bristles, followed by an irritation which lasts for some hours.

Injury Prevention

- Be careful when turning over rocks and corals.
- Wear gloves when fossicking on reefs.
- Avoid touching any very bristly worms.

Symptoms

- Contact with bristles initially produces a white, pinpoint rash, and the area may become red, swollen and numb over 10-30 minutes.
- The site of the sting may become stiff, swollen and painful over the next 24 hours.
- Itching or burning may last for a week and numbness over the area may persist for several weeks.
- General symptoms include increased pulse rate, palpitations, fainting and chest pain, causing the patient to feel very ill.

Treatment

- The bristles may be partly removed by applying adhesive tape to the area and peeling it off together with the bristles.
- Apply methylated spirits, calamine or other cooling lotions.
- Vinegar may assist in dissolving the bristles.
- Apply local anaesthetic ointment e.g. Lignocaine.
- Seek medical attention if general symptoms occur.

Worms that Bite

Eunice aphroditois

A very large reef species *Eunice aphroditois* may be more than a metre in length and 2 cm wide. It lives in crevices or under boulders on tropical reefs and extends some distance from its hole to forage on algae which it bites off with black chitinous jaws, which are large and strong enough to given an unpleasant bite to humans. It is widespread in warm Indo-Pacific waters and found on the north and west coasts of Western Australia.

Treatment

- Wash the area with antiseptic.
- Apply antibiotic ointment.

Crustaceans — Including Crabs, Prawns, etc.

Phylum Arthropoda
Class Crustacea

Figure 44: Some people develop allergies to crustaceans such as this aggressive blue manna crab *(Portunus pelagicus)* photographed in a sea-grass bed in Cockburn Sound.

Crustaceans, like insects, spiders and other arthropods, have jointed legs and a firm outer covering which is sometimes called a shell. This covering serves as a skeleton to which are attached the muscles which are the meat we eat. The skeletons of prawns are flexible, but crabs and rock-lobsters strengthen and harden their 'shells' with lime.

Crustaceans use their claws and smaller mouthparts to seize, tear and shred their food which, depending on their species, may be plants, animals or a mixture of both.

Allergies – a Simple Guide

The term 'allergy' is often used in discussions on marine illnesses, and it is necessary to explain what we mean by this word. An allergic reaction is one in which a person's body responds in an unusual and usually excessive way to a foreign substance, usually a protein, which causes little or no harm to most other people. The protein may be from plants or animals – particularly the marine creatures in this book.

Symptoms

Some unlucky people learn by experience that they cannot eat or even touch crayfish and/or prawns without developing an allergic reaction within a few hours.

- In some people the reaction may take the form of hay fever, with irritation of nose and eyes leading to sneezing and tears.
- Others may develop urticaria, when large areas of skin become red, swollen and itchy, sometimes with lumps or hives developing. A variant of this condition in which a part of the body such as a hand or part of the face swells is called angio-neurotic oedema. If swelling occurs inside the mouth and throat there can be a risk of suffocation.
- The most severe reaction is called anaphylactic shock when a person goes into a state of collapse with sweating, difficulty in breathing, severe disturbance to circulation involving falling blood pressure, and fading consciousness. This dangerous condition requires urgent medical treatment as life is in jeopardy.

However, it is important to realise that inflammation with pain, swelling, redness and impaired use can be caused by factors other than allergies. Such inflammation can result from injury, from foreign material entering the body or from irritant poisons contacting the skin, and may be complicated by secondary infection. Many marine animals produce toxins which cause reactions ranging from skin inflammation to the total collapse.

Anti-allergy Treatments

A characteristic of an allergic reaction is that it involves the release within the body of an inflammatory chemical called Histamine. The drugs known as antihistamines inhibit this release and can be very effective in treating allergic reactions.

- Urticaria is treated by the administration of an antihistamine by injection or by mouth for as long as the allergic reaction continues. This treatment shows its effectiveness by halting the red swelling of the skin and suppressing its intense itching. People often have antihistamine preparations which they use for other reasons such as treating hay fever or in preventing sea sickness. These preparations can be used if medical help is not immediately available.

• Really rapid relief from itching can follow the cautious subcutaneous injection of adrenalin which acts within seconds. Adrenalin is invaluable in treating life-threatening anaphylactic shock. It was formerly used in the treatment of other types of shock because it causes blood pressure to rise. However the modern treatment of shock is radically different.
• Cortisone, like adrenalin, is a natural body chemical and can be used in several ways. Creams and ointments of hydrocortisone and related chemicals can be used to dampen ordinary and also allergic inflammation. However they should not be used if the inflammation is due to an infection. These chemicals suppress the body's natural response to infection and so their use would ultimately lead to more tissue damage. For the same reason their use by mouth or injection should be prescribed with great caution. In the treatment of shock, however, cortisone may be injected intravenously in large dosages.

Outmoded forms of anti-allergy treatment include antihistamine skin creams which were found to cause a sensitization of their own with repeated use. A few topical antihistamine preparations such as eye drops are still sometimes used.

Poisonous Crabs

Family Xanthidae

Figure 45: Poisonous *Atergatis floridus* is commonly called the shawl crab because of its lacy colour pattern.

Of all crustaceans, only a handful of crab species are harmful or poisonous. Some poisoning has been traced to bad cooking or poor handling rather than to toxins in the crabs. However three species of xanthid or nut crabs are poisonous even after correct cooking and handling. One of the distinguishing features of nut crabs is the black tip of the finger of the claws. They differ from some other types of

Figure 46: This devil crab *(Zosimus aeneus)* was photographed in shallow water at Scott Reef north of Broome.

crabs in that they do not have any of their legs flattened to form swimming paddles like the blue manna *Portunus pelagicus* and the mud crab *Scylla serrata.*

Of the three toxic species, the shawl crab *(Atergatis floridus)* (Figure 45) and devil crab *(Zosimus aeneus)* (Figure 46) are common on the tropical shallow reefs of northern Australia, while the other, the rough crab *(Platypodia granulosa),* has not yet been found in Western Australia.

The toxicity of these species from Western Australia has not been tested and no case of severe sickness or death after eating crabs has been reported. The most dangerously poisonous species is *Zosimus aeneus* since it is the largest and has the greatest toxicity. Half a gram of the claw muscle of this species may contain sufficient poison to kill a person. The poison is thought to be saxitoxin, an extremely powerful venom. About 5,000 mice can be killed by only 1 mg of saxitoxin, which is formed inside the cells of minute planktonic algae called dinoflagellates. These are filtered from the water and eaten by molluscs such as mussels and clams (see page 99) which accumulate the toxin. In turn, these molluscs are presumed to be eaten by these crabs. Neither the molluscs nor the crabs are sensitive to this toxin, but it paralyses those who eat them, and has been named paralytic shellfish poison.

This poison is not destroyed by heat, which is unusual for marine toxins. It is a neuro-toxin and prevents nerves – and so muscles – from functioning normally. Asphyxiation may occur if the muscles responsible for breathing are affected.

Poisoning Prevention

- Avoid eating any crabs with black-tipped claws.
- Do not ever eat any of the species known to be toxic.

Symptoms

- Within 15 minutes to a few hours the lips and tongue become numb and the limbs weak. This is followed by severe vomiting and diarrhoea.
- Dizziness, drowsiness, muscle paralysis, burning in the throat and stomach, respiratory failure and coma.
- Death may occur within four to six hours after ingestion.
- The severity of symptoms depend on the species of crab and the amount eaten. In 530 cases reported in the literature there have been 37 deaths.

Treatment

- Seek expert medical attention as quickly as possible. The treatment is completely directed at keeping respiration going (see page 94).

Molluscs — Including Cone Shells, Octopus, etc.

Phylum Mollusca

Figure 47: Trying to right itself, this false ear shell *(Granata imbricata)* extends its muscular foot and exudes a milky fluid which is thought to deter predators.

Most of us are familiar with at least some molluscs. We recognise dead shells washed up on the beach, oysters squids and scallops which we eat, or even snails and slugs which are pests in our gardens. However it is not easy to see why these animals would be grouped together as molluscs.

Most molluscs have shells to protect their soft parts, though squids, for instance, have only a thin internal shell for muscle support and some slugs don't have a shell at all.

Among the most noticeable external features are the sense organs such as eyes and tentacles with which the mollusc sees, feels and tastes its environment. Muscular organs like the crawling foot of a snail react to the environment by causing movement, perhaps towards food or away from danger. In a bivalve like a mussel the muscles close the shell until conditions are favourable again. Attached to the muscles are the internal organs of digestion, excretion and reproduction, as well as the nerves and the heart and blood passages.

Covering all these organs is a skin called the mantle which forms the shell. It typically forms a flap or pocket behind the head which may shelter gills, and/or function as a lung.

There is a great variety of body forms and life styles amongst molluscs ranging from fixed, heavy-shelled giant clams to streamlined thin-shelled swimming scallops; from abalone which have large gills and which live in the sea to air-breathing ramshorn snails in freshwater ponds; from slow-moving, weed-eating limpets to swift squid which may even leap out of the water when chasing fish. Some snail-like gastropod molluscs such as periwinkles scrape their food from rocks with a rasp-like ribbon of minute teeth called a radula. Others are carnivores, and their teeth and jaws are specially shaped for catching their prey. On the other hand, bivalves with their two-part shells generally sieve tiny food particles from the water and do not have either jaws or a radula.

There are two ways in which certain marine molluscs can harm and even kill humans. Some such as cone shells and octopus cause injury by using the mechanisms with which they normally obtain food. Others may cause illness when eaten.

Some molluscs use the radular teeth and perhaps also the jaws to wound prey. The wound itself is usually not dangerous – particularly to larger animals including humans. However, if digestive juices and/or toxic fluids are injected through the wound, more serious consequences can follow.

Figure 48: The snout and the stalked eye of this anemone cone *(Conus anemone)* are below the long tubular siphon and above the more brightly coloured foot.

Cone Shells

Class Gastropoda
Family Conidae

Figure 49: The colour pattern of some living cone shells is hidden beneath a horny outer layer or periostracum which, in this geographer cone *(Conus geographus)*, bears bunches of hairs.

Cone shells, as their name indicates, have a cone-shaped shell which, in a few species, has a flat end though in most species that hind end is raised or pointed. The sides of the cone shell may be straight or curved and convex. The shell surface can be smooth, granular or ridged and is ornamented with a colour pattern characteristic of the species. This pattern is sometimes hidden by a horny outer layer of periostracum which might be thick and even hairy. The shell itself can be thick and heavy or thin, lightweight and more suitable for a fast-moving animal.

Each species of cone shell has a characteristic size, shape, shell-texture and colour-pattern, which usually varies a little between individuals of the one species.

One of the most obvious features of an active cone's body is its siphon. This is a split tube which can be extended and contracted, and which protrudes through a notch in the narrow anterior end of the shell just above the cone's head. It is like a snorkel, and it is through this siphon that the cone draws a water current into its mantle cavity. This current of water passes through the gills for respiration and also over an organ which detects the scent of animals including those on which it feeds.

Figure 50: Three teeth taken from the tooth sac inside a striated cone *(Conus striatus).*

The Harpoon

Most cones prey upon worms or molluscs including other cones. Some, however, kill and eat small fish such as reef blennies. Like most predators of mobile animals, a cone shell has a very efficient sensory and nervous system, and can move rapidly. Sampling the water in different directions with its flexible siphon, it is able to home in on its prey by sensing the chemicals spreading from it through the water, even when the prey is moving.

Cones which catch slow-moving worms or molluscs seem to approach their prey fairly slowly. When near, the cone extends its proboscis, which is normally kept tucked inside the snout, seeks out an appropriate area of the prey, and then suddenly strikes, plunging a small harpoon into the prey's body. Some cones are reported to shoot their harpoon like the dart of a blowpipe. This harpoon is a hollow tube with a barbed point. It is really a single tooth which has been formed, and then stored, in a small sac which opens into the mouth cavity. The teeth of cones, unlike those of most other gastropods, are not joined to form a rasping radula but are quite separate, though they lie in bundles inside a storage sac. When needed, a tooth is passed from the tooth sac through the mouth cavity and along to the tip of the proboscis. As the cone's tooth pierces the body of the prey, venom is injected into the wound. This venom has come from the venom gland through the mouth cavity, up to the tip of the proboscis and into the hollow harpoon. The victim is rapidly paralysed by one or more teeth and the venom they contain, and the muscular proboscis then spreads around and engulfs its immobile prey.

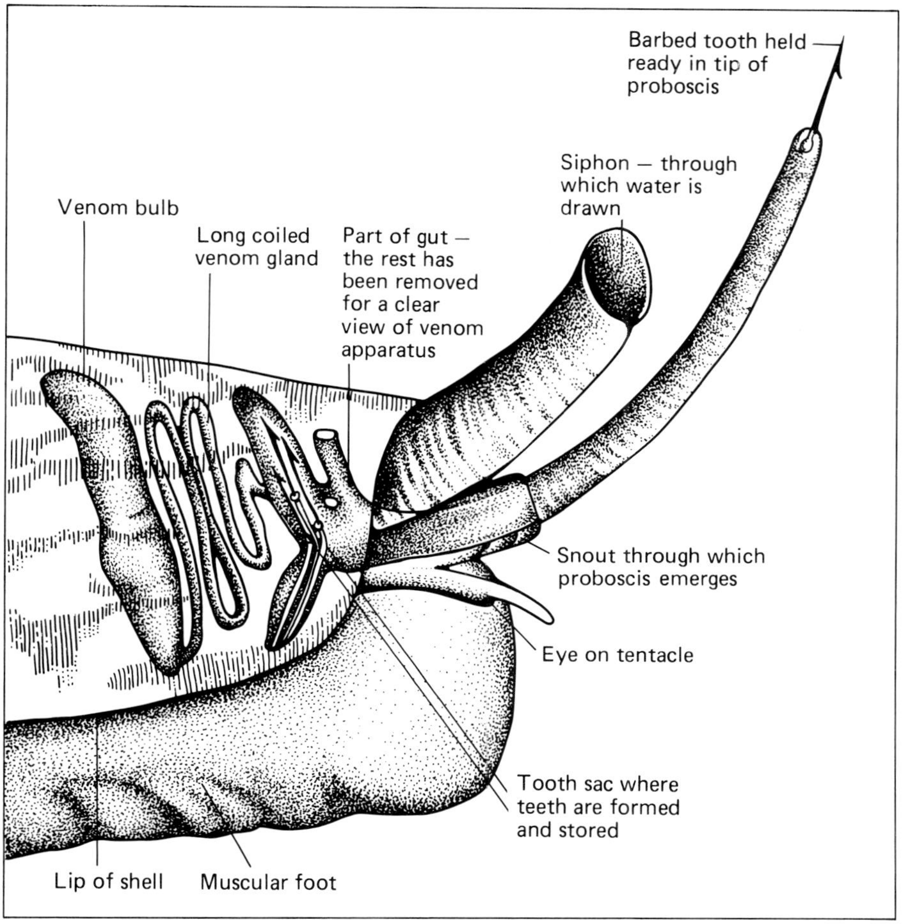

Figure 51: Diagram of the organs in a cone's body which are associated with prey capture and feeding.

Fish-eating cones need to act quickly to catch their rapidly-moving prey. Even when buried under the sand, but using its protruding snorkel-like siphon, the cone can track the fish as it swims nearby. It then leaps out of the sand to plunge a tooth into the fish's body. On other occasions fish-eating cones are known to creep up on their prey with slow, gliding movements. In aquaria some cones will eat pieces of fish flesh without using the venom apparatus. They simply open out the proboscis and move up to envelope the food.

Sometimes the prey is quite large compared with the size of the cone, and cones have been known to swallow prey of more than half their own size. The cone, with its body greatly distended, retreats to shelter under the sand or into a crevice to digest its meal.

Cones Dangerous to Man

The behaviour of only a few species of cones has been thoroughly investigated. Most species seem to attack their prey only at night but their feeding activity is governed more by hunger than by the time of day. The potency of cone shell venom seems to vary not only between species, but also between individuals within a species. It can also vary from one time to another within a single animal. Venoms which have been investigated consist of more than one type of toxin and these might mature at different rates.

Many cone shell injuries have been inflicted when a live cone is being carried in a pocket or fold of clothing or in a bag of thin material. The stinging reaction can, apparently, be triggered off by pressure, rubbing or close confinement, particularly when a cone shell is crushed among other animals. The cone then acts like an animal backed into a corner, and strikes at anything within reach.

The species of cone shells apparently most dangerous to man are those which normally feed upon fish. This conclusion is supported by experimental work on the components and action of the venoms of cone shells. Most fish-eating cones have long teeth with sharp cutting blades, perhaps to ensure deeper penetration of the venom. The venom injected through these long harpoons causes numbness, swelling and sometimes pain around the wound. General numbness and paralysis of skeletal muscles, which are those attached to bones, can rapidly follow. Though the venom may affect heart muscles and the central nervous system, the most important effect is on the muscles of the chest wall and of the diaphragm. Paralysis of these muscles can cause breathing difficulty or even stoppage.

Figure 52: The head of a striated cone showing a tooth held in its mouth.

Cones which eat worms have complex venoms containing many chemicals including some protein-digesting enzymes. It might be these enzymes which cause pain, cell distintegration, bruising, swelling and bleeding when injected into larger animals including humans. The venom of these species does not seem to cause paralysis, even of the worms on which they prey, though it might slow them down. The teeth of such cones are relatively short and robust with large knobs at their bases around which the tip of the proboscis apparently contracts, to hold the tooth firmly.

There has been some debate about the danger to man from mollusc-eating cones. Experiments now suggest that this group is unlikely to seriously injure either fishes or humans. Their venoms do not appear to paralyse human muscles, though some affect the central nervous system, occasionally causing other severe symptoms. The teeth of these cones are relatively long with slightly enlarged bases and they lack cutting blades near the point.

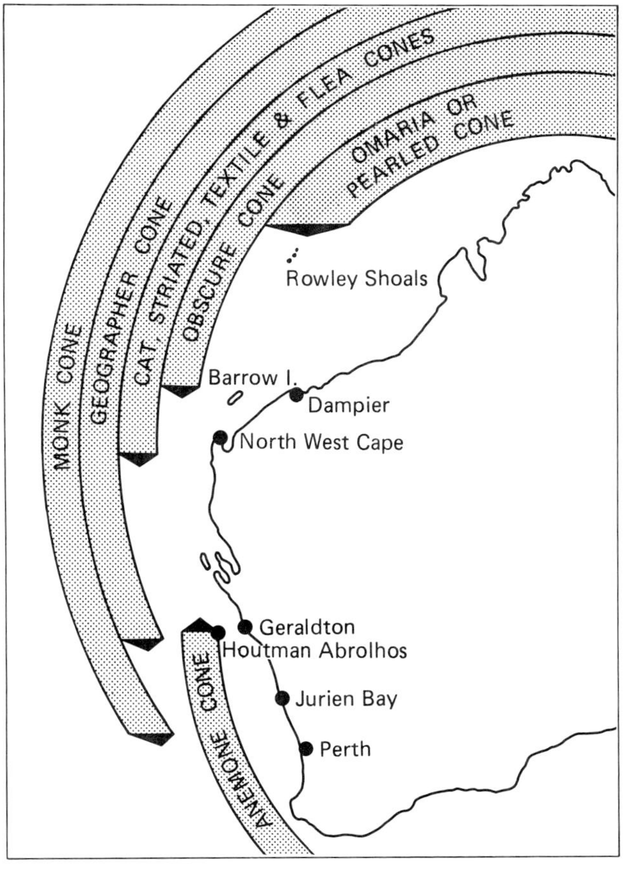

Figure 53: The distribution along the Western Australian coast of species of cones known or thought to be dangerous.

Figure 54: C. geographus.

Geographer Cone *(Conus geographus)*

The Geographer Cone is known to be responsible for 12 human fatalities, as well as many serious injuries. Never touch this cone, and take great care if you encounter one.

Its shell is thin, smooth and light-weight. It has a low spire, and is shaped more like a swollen cylinder than a cone. The outside is pale blue or pale pink with a fairly strong pattern of red-brown lines forming a network to enclose small triangular patches. This is called a tent pattern. Around the shell there are irregular bands of brownish blotches, formed by a denser or filled-in tent pattern. Inside, the shell is white though some of the exterior colour pattern may show through the thin shell which can be up to 13 cm long.

The Geographer Cone is widely distributed through the tropical regions of the Indian and Pacific Oceans, being found in Western Australia as far south as the Houtman Abrolhos off Geraldton. It eats fish and is not often seen alive because it hides in the sand under coral slabs and rock ledges during the day. Aquarium studies have shown that this cone can recognise pieces of fish flesh which it can eat without using its poison apparatus (U. Weinreich, pers. comm.).

The venom of the Geographer Cone contains a number of toxins, some acting directly on human skeletal muscles and one affecting the central nervous system. If paralysis occurs the muscles become limp. Severe pain and swelling may occur.

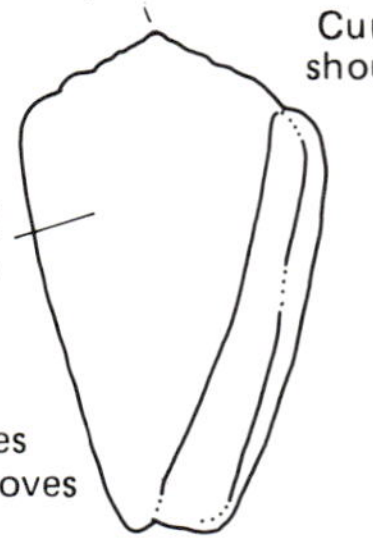

Figure 55: C. catus.

Cat Cone *(Conus catus)*

The shell of the Cat Cone is rather small and stout, and is quite heavy for its size. The sides of the body whorl are slightly curved, particularly near the raised, straight-sided spire. The texture of the body whorl is roughened by grooves and fine nodular ridges which run around the shell. The colour of the shell is brown with irregular broken patches of white or blue which tend to form radial bands along the length of the body whorl. These intersect two bands of darker patches running around the body whorl. Near the anterior end there are white spots which in some shells form narrow radial bands across the ridges. The shell can be up to 5 cm long.

The Cat Cone is abundant on the tropical shores of the Indian and Pacific Oceans and as far south as North West Cape in Western Australia. It hides in crevices in wave-swept intertidal reefs and is said to feed on fish. It has inflicted only minor injuries to humans, but despite its small size it should be treated with caution.

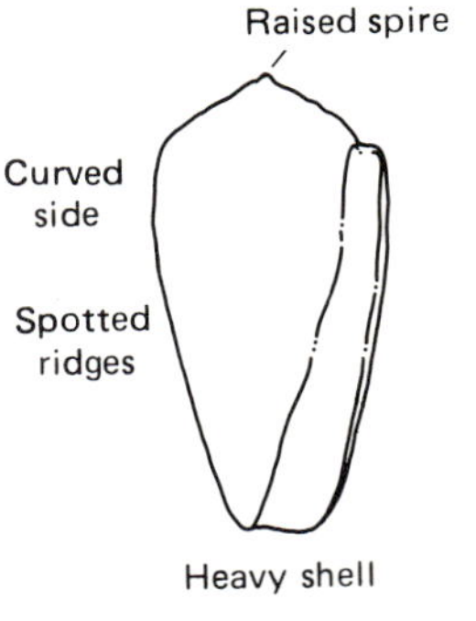

Figure 56: C. monachus.

Monk Cone *(Conus monachus)*

This shell is up to 6 cm long and resembles the Cat Cone in shape and texture. The colour pattern is variable, but is usually of white and green or brown splashes of colour on a blue-grey background. The white spiral ridges are usually patterned with fine spots of reddish-brown. In some specimens this brown colour is dominant, broken only by spots of white, while in others there is about the same amount of the two colours on the ridges. *Conus achatinus* is a name given to a variant of this species.

The Monk Cone is quite common in Malaysia, the Philippines and Indonesia. It is found on the northern coasts of Australia and south to about Jurien Bay on the west coast. Like most other cones it stays in sheltered places during the day. It is thought to eat fish and has teeth shaped like those of other fish-eaters. Although it has not yet been recorded as injuring humans, it is potentially dangerous as experiments show that its venom can paralyse skeletal muscles of mammals.

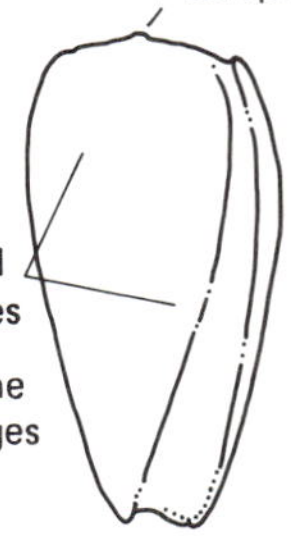

Figure 57: C. striatus.

Striated Cone *(Conus striatus)*

This cone has a large shell which is rather thick and heavy and can be up to 10 cm long. The sides of the body whorl are curved, particularly near the flat spire. The shell is usually pale pink with a pattern of large irregular blotches of purple-brown. Darker, more solid blotches tend to form two broken irregular bands going around the central part of the body whorl.

The Striated Cone is widespread in the Indian and Pacific Oceans, but has only rarely been found in Western Australia around North West Cape, Onslow, Dampier and offshore reefs such as Rowley Shoals, where it hides in sand under coral slabs, boulders and ledges during the day. It eats fish and should be regarded as potentially dangerous, though no serious injuries have been recorded. Tests have shown that its venom is less potent than that of the Geographer Cone, with at least some of the toxins being of a different type, paralysing skeletal muscle in a contracted state. Studies on the production of venom from its relatively small venom gland suggest that not enough could be produced to kill a person.

Figure 58: C. obscurus.

Obscure Cone *(Conus obscurus)*

This small cone is like a miniature Geographer Cone, having a thin, smooth, almost cylindrical violet shell of up to 4 cm in length, with a brown pattern in indefinite spiral bands. However it has a higher spire than the Geographer Cone and has spiral rows of small white and brown patches instead of a triangular pattern. Inside, the shell is violet clouded by the colour pattern showing through.

It shelters under coral, and is usually buried in sand. It lives in tropical areas of the Indian and West-Pacific Oceans, and in Western Australia has been found as far south as Barrow Island.

Although it feeds on fish, the injuries reputedly caused by it have only affected the area around the site of the puncture wound and have not been severe or long-lasting. It is possible that this small cone could not secrete enough venom to cause serious injury to humans, but it should be treated with caution.

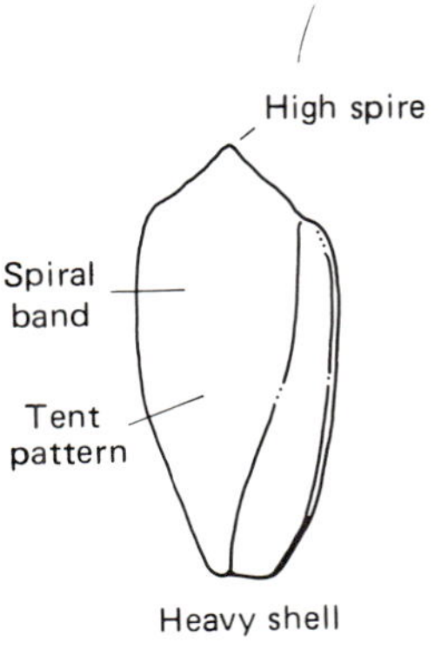

Figure 59: C. textile.

Textile or Cloth-of-Gold Cone (*Conus textile*)

This has a fairly smooth, glossy shell, which is rather heavy for its size. Its shape varies from slender and fairly straight-sided to squat with bulging sides. It has a rather high spire with straight or concave sides. The shell is decorated with a tent pattern, with triangular patches of the white background outlined in dark brown. Broken areas of light brown with irregular dark brown radial lines form spiral bands running around the body whorl and the spire. The interior of the shell is white, and the shell is up to 10 cm long.

The Textile Cone lives in tropical waters of the Indian and Pacific Oceans, and is fairly common on the Western Australian coast south to about Point Coates. During the day it hides under stones, coral slabs and ledges, and probably comes out to feed at night. It feeds on gastropods, molluscs including other cones, and can even kill an attacking octopus. It may severely injure humans, and has been reported as causing two deaths, although the identifications were possibly unreliable. Recent experimental work on its venom shows that it does not directly paralyse skeletal muscle. However it should be treated with great caution, as there is evidence that the venom might affect the central nervous system.

Aquarium studies indicate that this cone will sometimes feed upon worms as well as gastropods, and that it can fire its tooth from a distance of a centimetre or two with enough force to penetrate the horny operculum of a periwinkle (Scrimgeour, 1985).

Figure 60: C. omaria.

Pearled Cone *(Conus omaria)*

This cone resembles the Textile Cone but its shell is smaller, being less than 8 cm long, and it tends to be more slender with a lower spire and a more sharply-angled shoulder. The white background colour is broken into large well-defined but irregular triangular areas. Large patches of brown, delicately spotted with white, form two broken but wide spiral bands around the body whorl. The white shell interior is seen through the broad aperture.

The Pearled Cone is found through tropical areas of the Indo-West Pacific region, including the north coast of Western Australia. It is rarely seen, though it lives in shallow coral reef areas as well as in deeper water.

The venom of this mollusc-eating cone has been recorded to cause severe symptoms, and so, like the Textile Cone, should be handled carefully.

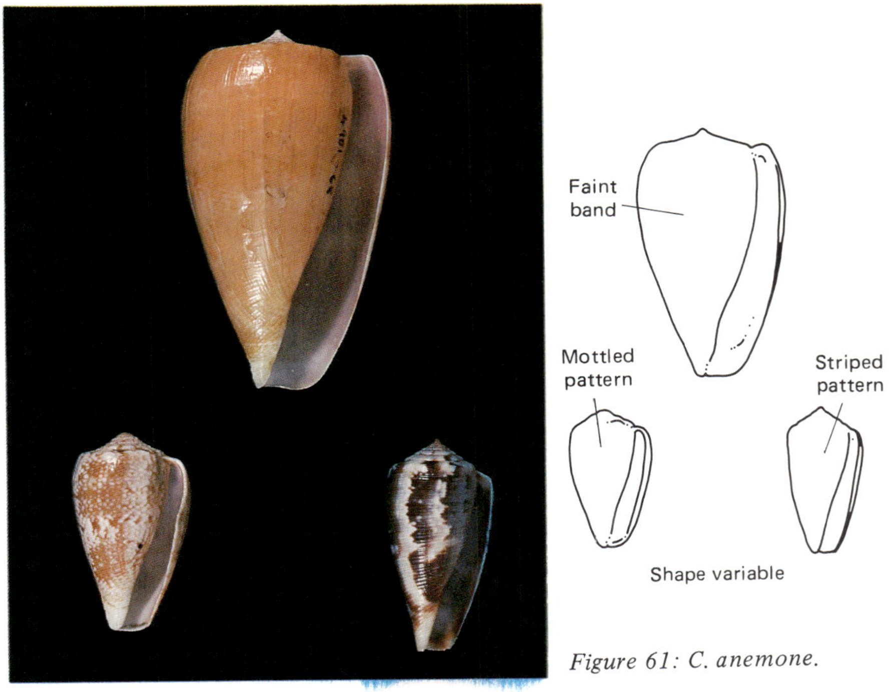

Figure 61: C. anemone.

Anemone Cone *(Conus anemone)*

This common cone is extremely variable in shape, pattern and colour. The soft parts are usually in shades of pink, while the background colour of the shell can range from almost white through yellow and pink to blue. Its mottled pattern of red, brown or purple can be light or heavy, and forms two or three broken and sometimes indistinct bands around the body whorl. The shell can be moderately heavy, or lightweight and fragile, and thinner shells usually are more rounded than thick shells. Apparently mature shells range from about 2.5 to 6.5 cm long. This cone occurs along the southern coasts of Australia, and along the west coast to just north of Geraldton. It is found under rocks, and sometimes in sand in intertidal areas and in deeper water.

Although the Anemone Cone is thought to be a worm eater and so would not be expected to be included among dangerous cones, it has caused moderately painful but localised swelling in at least three cases. It is included here because it is common in the more highly populated areas of Australia.

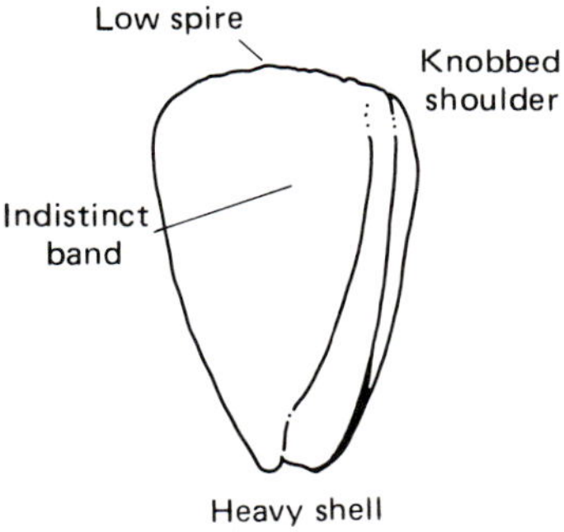

Figure 62: C. pulicarius.

Flea Cone *(Conus pulicarius)*

This cone shell eats worms and has a fairly heavy cone-shaped shell with a low, pointed spire. The sides of the body whorl are slightly rounded anterior to the shoulder which is knobbed. Black spots are scattered over the white shell. It has two indistinct spiral bands where the spots are larger and more dense. The interior of the shell through the fairly narrow aperture is pale, usually white.

The Flea Cone is common in the tropical Indo-West Pacific region and its distribution in Western Australia extends to an area south of North West Cape. This cone is usually buried in sand, as are the worms on which it feeds.

Like other worm-eaters which injure humans, this cone injects venom causing bleeding and pain around the sting. These effects are possibly due to protein-digesting enzymes in the venom.

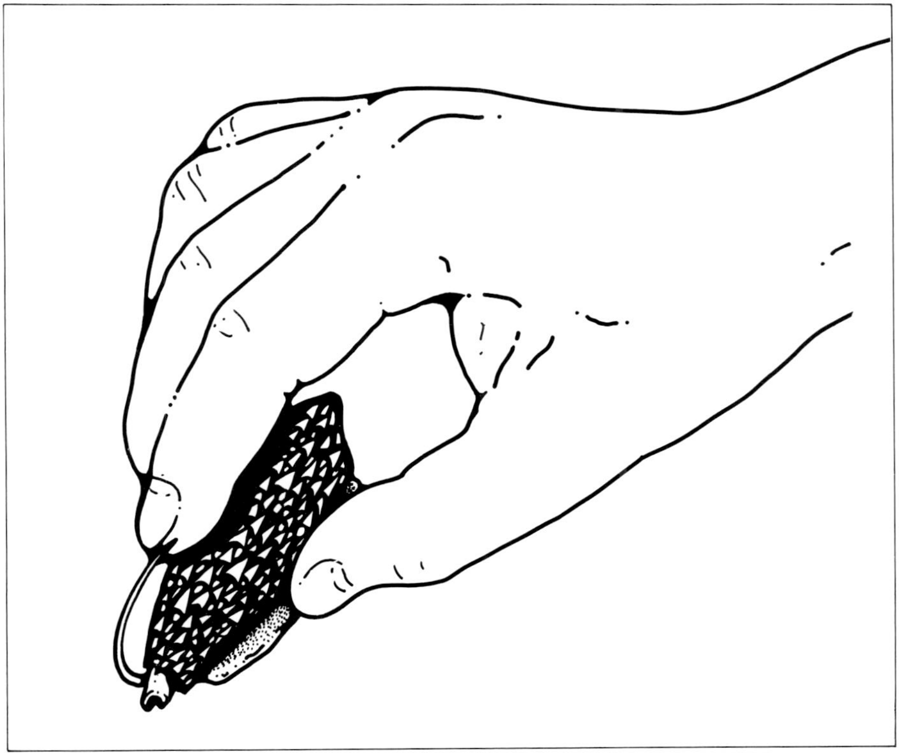

Figure 63: Great care is needed when handling a live cone as its proboscis can quickly become very long.

Injury Prevention

● All living cones should be regarded as potentially dangerous, and so should be handled as little as possible. Collectors are advised to wear heavy gloves, or use a knife, stick or bent wire when dislodging cones from under rocks or within crevices. Live cones should be rolled over a couple of times so that the animals retreat into their shells. They can then be picked up but should always be held at the widest part of the shell near the spire. However cones are very agile and can twist their bodies through almost 180° as they extend out of the shell. The proboscis can elongate right back to the shell's spire, so cones should be quickly placed in a collecting bag or pail. They should never be held in the hand, put in a pocket or placed inside a wetsuit. They can sting through clothing.

Symptoms

● Although only a few species of cone shells are known to inflict serious injury or cause death, nobody knows how many other species may be dangerous to humans. There also seems to be a great variation in people's susceptibility to the venom. More research is needed on the toxicity of cone venom, the factors which affect feeding in cones, and why and when they strike. Do not take risks with any cones, particularly *Conus geographus,* which is extremely dangerous.

- The different species of venomous cone shells cause a variety of symptoms, but the most important effect is on the breathing process. There may be an obvious sharp pain where the victim is stung, and the area around it may seem to burn and may swell and become numb. The more serious symptoms are paralysis which may become general, weakness, clumsiness, difficulty in swallowing and speaking and the development of vision problems. The most important symptoms are increasing difficulty with, and perhaps ultimately failure of, the breathing process. The symptoms can occur quickly, progress over several hours, reach a peak and then go away.

Treatment

- This is directed at keeping the victim alive through the period that the venom has its greatest effect. The methods are similar to those involved with the treatment of snake bite, except that, as yet, there is no cone shell anti-venom available.
- The area bitten – usually a hand – should be bound with a pressure bandage and immobilised in a sling. This is aimed at delaying the spread of the venom through the rest of the body. The pressure bandage, as in a case of snake bite, should not be excessively tight. It is important to avoid any risk to the vitality of the limb. There should thus be no need to take off the pressure bandage until the patient has been brought to medical attention. When the bandage is taken off there will be a very real risk that the sudden release of the venom will have a drastic effect on the nervous system.
- A bite to the torso should have a pressure pad as well as a pressure bandage applied to it. Not only should the actual site be immobilised but also the whole body of the victim, who should be carried if there is a need to move.
- If the victim loses the ability to swallow, particularly if breathing becomes difficult, treatment by artificial respiration becomes of paramount importance. This artificial respiration must be kept up until the effects of the venom wear off. If the heartbeat should fade, cardiac massage must also be added to artificial respiration.
- It is important to remember that although the victim may be paralysed he is still conscious and can see and hear. It is vitally important to realise that he can understand and will remember what is said to him, even if he can't answer. Symptoms and the treatment should be carefully explained to reduce stress and prevent panic. The victim should never be left alone. Not only can paralysis cause breathing failure but vomited material may be inhaled and cause choking.
- Anyone likely to vomit should be put in the Coma Position – lying on one side – so that choking will not occur.

Octopus, Squid and Cuttlefish

Class Cephalopoda

Figure 64: This solitary reef-dwelling cuttlefish *(Sepia apama)* is curious and can become aggressive to divers. Bites may become inflamed.

Cephalopod molluscs are active predators with very efficient sense organs and complex nervous systems which allow them to react quickly. They locate their prey mainly by sight, possessing eyes like ours which can see objects. In this way they are unlike other molluscs such as gastropods which can only distinguish between light and shade. The large cephalopod brain not only stores complex inherited behaviour patterns, but can also learn and remember from experience.

Squid, cuttlefish and octopus have a soft but muscular body without a protective shell. Around the mouth are eight arms each with one or two rows of suckers. Inside the mouth is a pair of jaws shaped like a parrot's beak and a many-toothed radular ribbon.

Squid and cuttlefish also have a pair of tentacles which are used to catch hold of their prey. Their suckers are strengthened by a horny-toothed rim. In some squid the suckers on the tentacles are replaced by sharp, horny hooks. Most species can swim swiftly. Normally they use fins on the sides or at the end of the body or mantle. The muscles of the mantle and the fins pull against an internal shell which, in squids, is horny, transparent and flexible, but in cuttlefish is the rigid, chalky cuttlebone often fed to birds. When needing an extra burst of speed to catch prey or escape from danger, these cephalopods use jet-propulsion, forcing water out of the mantle cavity under pressure. Squids and cuttlefish may swim in schools or alone, and they hunt fish crustaceans and other cephalopods in the open ocean or in shallower coastal waters.

Figure 65: Most octopus creep over the substrate changing their colour pattern and sometimes their skin texture to merge into the background.

Octopus lack the internal shell, the two long tentacles and the toothed suckers of squid and cuttlefish, and most of them don't have fins. Instead, they creep across rocks and sand and through weed searching for crabs and molluscs for food. On occasions, they do swim by jet-propulsion but usually only for short periods as most tire quickly. When hunting at night or in dark caves, octopus probe with their eight muscular but sensitive arms, picking up the scent of their food with sense organs on the skin of their suckers which they also use to grasp and hold their prey.

Different octopus species use various methods to subdue and kill food animals. Some bore through the shell of well-protected molluscs using the wide rasp-like radula and, in some cases, fine accessory teeth within the mouth. Others are reported to use their jaws to bite animals such as crabs and to inject saliva into the wound. But many species hold the victim within an umbrella formed by the webbed arms. Saliva, squirted into this confined space, is apparently absorbed into the tissues of the victim, paralysing and killing it.

Octopus have two sets of salivary glands. One pair in the 'neck' region behind the eyes produces a thick venomous saliva. These glands are particularly large in blue-ringed octopus. Saliva from a smaller pair of glands near the mouth located at the centre of the tentacle ring may contain a spreading agent to thin the venom and perhaps to aid its penetration. The saliva of many octopus species is digestive and loosens the attachment of the crab's or mollusc's muscles to its shell. The octopus then pulls the flesh from even the finest cavities so that a clean but almost unbroken shell remains.

Octopus escape from enemies such as large fish and seals by rapid jet-propelled swimming. They can also control their colour pattern and skin texture for camouflage. When threatened or when excited

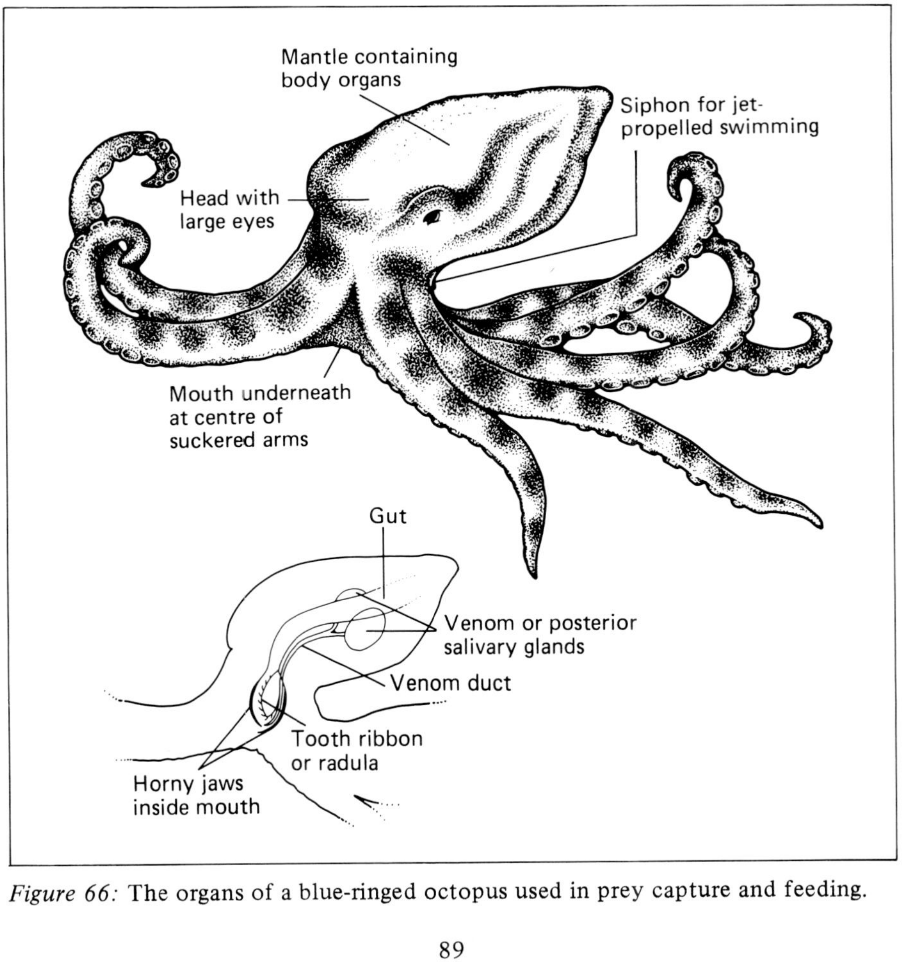

Figure 66: The organs of a blue-ringed octopus used in prey capture and feeding.

by the presence of prey or sometimes of another octopus, the colour change can be dramatic and the pattern is usually quite characteristic of the species. The response of a blue-ringed octopus is particularly startling. Rings or bands of blue appear, highlighted by expanded black rings or bands which surround them. As rapidly as this colour pattern appears, it can change back to the camouflage pattern of mottled yellow-brown with the blue bands or rings almost invisible.

Most octopuses and other cephalopods possess an ink gland and puff out a cloud of dark pigment into the water when disturbed. This cloud, appearing suddenly from the mantle funnel, startles a pursuer and momentarily hides the cephalopod which instantly becomes paler and swims away.

Injury Prevention

- To avoid being bitten when holding a cephalopod, grasp the body behind, but close to the head and eyes. The adhesion of its suckers might feel uncomfortable but causes no harm except, perhaps, a very slight abrasion from the serrated sucker rings.
- Underwater, panic can cause much more damage than the bite of a cephalopod – even a venomous species – could ever do. If a cuttlefish, attracted by the shining glass and metal, grasps a diver's face-mask, the rules governing safe ascent to the surface should be rigidly followed. The knowledge that frightened or injured cephalopods will spurt 'ink' and grasp onto anything within reach should help a diver to avoid the panic which could lead to the 'bends' or drowning.

Symptoms

- The bites of squid, cuttlefish and octopus produce symptoms which are related to the amounts and types of chemicals present in the saliva. In most species investigated these consist of digestive substances as well as toxins. The wound into which these substances have entered might bleed freely, though there is usually little pain. Redness and swelling around the wound may develop, even some hours later. Some tingling, numbness and even 'light-headedness' and weakness may occur.

Treatment

- The symptoms caused by cephalopod bites (other than those of blue-ringed octopus) usually pass off after a few hours and the wounds usually heal without any complication.
- Clean the wound and close it.
- Large wounds need suturing.
- Apply Tetanus booster.

Blue-ringed Octopus

Family Octopodidae

As mentioned above, blue-ringed octopus of the genus *Hapalochlaena* are noted for the vivid colour pattern which develops when they are excited. They are also distinguished by being relatively small octopus and by having a fairly deep web between their relatively short arms. Though blue-ringed octopus are said to have ink glands for only a short period after they hatch, one tropical species has been observed to produce a small quantity of brown ink when extremely irritated.

Observations on Australian blue-ringed octopus indicate that they use a highly toxic venom to quickly kill their prey – generally, or perhaps exclusively, small crustaceans such as crabs. This venom has caused human deaths and is said to be more potent than the venom of the most deadly snake.

Figure 67: Except when excited, southern blue-ringed octopus are not colourful and may be drab brown.

Southern Blue-ringed Octopus *(Hapalochlaena maculosa)*

This octopus is common around the southern coast of Australia north to about Shark Bay in Western Australia. It is not often seen, however, because it is small, rarely exceeding 12 cm across the arms, because of its powers of camouflage and because of the speed with which it can move from one place of concealment to the next. It is found sheltering within dead gastropod shells, clumps of mussels and discarded bottles and cans. It frequents algal-covered boulder piles exposed to pounding waves as well as sheltered sea-grass beds, from the intertidal areas to depths of at least 10 metres.

The female often inhabits empty, hinged bivalve shells when laying and brooding eggs. Her suckered arms hold the shell valves together to protect herself and the eggs she guards. These eggs are relatively large

Figure 68: Hapalochlaena maculosa from the southern coast of Western Australia flashes its brilliant blue rings and bands when disturbed.

for such a small octopus and are arranged in clusters attached to one another but not to the substrate. About 100-150 egg capsules, each approximately 7 mm long, are held in the groove between the mother's body and the webbed base of her uplifted arms.

As in most cephalopods studied, the female does not feed while brooding her eggs which take about two months to develop and hatch. The young octopus crawl on the bottom almost immediately after hatching, staying with their mother for several days before moving off to fend for themselves. Still not eating, the mother might live for a few more weeks before dying at the age of about one year. It is presumed that the male of this species also dies after breeding, as happens in other species of octopus studied.

Newly hatched southern blue-ringed octopus are anti-social to the point of eating each other, but by the time they are a month old they can catch and kill small crabs which seem to be their preferred food.

There is some doubt as to how these blue-ringed octopus inject venom into humans or, for that matter, into crabs. Some reports indicate that no incisions could be found on the skin of victims, but only faint bruising which rapidly faded. Other reports describe lacerations 5 mm deep probably caused by the octopus' beak.

The venom has been intensively investigated, and contains two toxic substances. Maculotoxin (said by some workers to be identical with the poison tetrodotoxin from puffer fish) is apparently not toxic to crabs, but is toxic to fish. It therefore must be a venom used in defence if these octopus do not normally feed upon fish. The other toxin, hapalotoxin, is effective on crabs, and is similar to toxins produced by other crustacean-eating cephalopods. Both maculotoxin and hapalotoxin affect the transmission of impulses between nerves and skeletal muscles in mammals. Their relatively small molecules do not produce an antigen response and so, at present, it seems that no immunisation method can be developed against these venoms.

Figure 69: The large northern blue-ringed octopus can be recognised by its large open rings and smoother skin. See also cover plate.

Northern Blue-ringed Octopus

The larger of the two northern coast species has generally been known as *Hapalochlaena lunulata,* but is now thought to be an unnamed species. It is larger than the southern species, growing to about 20 cm when measured across its extended arms. It is found south to about the Houtman Abrolhos, and lives under rocks and dead coral on reef flats and also in tidal pools in more muddy areas. Like related species it is not often seen because of its retiring habits and ability to camouflage.

One human death, which occurred at Darwin in the Northern Territory, has been attributed to this species. Only a very small puncture wound was found on the victim's skin and there was no development of localised swelling or pain around the bite.

A smaller species is now considered to be the true *Hapalochlaena lunulata.* It is very similar to the southern blue-ringed octopus in both size and form but has a slightly different arrangement of the blue rings on its body. This species was first recorded from islands off eastern Papua New Guinea. It occurs in eastern Indonesia as well as along the north coasts of Australia south to at least Broome in Western Australia, and inhabits shallow intertidal pools as well as turbid deeper waters around reefs and jetty piles.

Though nothing is known of the effects of its bite on humans, an aquarium specimen of this species killed a large sea snake within a minute of leaping onto its back. This could indicate that the feeding venom is also used as a defence against natural predators.

At least three species of blue-ringed octopus are known to live off northern Australia.

Injury Prevention

- Don't catch or hold blue-ringed octopus, or allow one to crawl on the skin. They are more likely to bite if lifted out of the water.
- Because these octopus may inhabit empty shells, bottles, etc. don't carry those collected underwater in the hand, in a pocket or tucked inside a wetsuit. When collecting or sorting mussels take care as octopus often live among them. When hauling crab nets or rock-lobster pots remember that blue-ringed octopus often shelter in them. A cray fisherman was bitten when, unknowingly, he tried to brush off an octopus crawling on his leg.

Symptoms

- The bite of a blue-ringed octopus is rarely felt and does not cause pain or swelling, though some bleeding may occur. Often it is so small, that it cannot be located. The first indication that a bite has occurred may be a feeling of numbness and an increasing difficulty to speak and to see clearly.
- Paralysis may develop and respiratory failure is a real danger.

Treatment

- It is important to remember that the patient should never be left unattended.
- If the site of the bite is known then pressure should be applied with a pressure bandage around the limb or a pressure pad and firm bandage if the bite is on the torso. The limb, and indeed the whole body should be immobilised as far as possible.
- Most importantly, the breathing of the victim should be carefully watched, as respiratory failure would cause death. If the breathing starts to fade, then mouth-to-mouth or an equivalent artificial form of respiration must be maintained until the patient is in a medical centre which can take over ventilation. Life support may be needed for up to 24 hours.

• A particular problem is caused by vomiting, as the vomited material may lodge at the back of the paralysed throat causing choking. Constant checking of the airway to the lungs is essential and the clearing of any obstruction should be carried out immediately, using the fingers if necessary. A victim likely to vomit should be put in the Coma Position – lying on one side – so that choking will not occur.

• The care of the patient is demanding and requires constant watchfulness. The patient may be totally paralysed and yet fully conscious, hearing and remembering everything. The kindest and most efficient way to cope with the distress and even panic that is inevitable, is to explain to him what is happening and to reassure him that he will not be left alone.

Poisonous Molluscs

Figure 70: Many nudibranchs like this *Phyllidia verrucosa* from Scott Reef are distasteful and even poisonous to fish, some other invertebrates and probably humans.

Most molluscs are edible and many are widely used for food being well-flavoured and nourishing. Some, such as oysters, may be regarded as luxuries.

However, some unfortunate people develop allergies to one or more types of molluscan foods which others eat without problems. A few molluscs are occasionally contaminated by pollutants or by naturally occuring poisons and a very few molluscan species contain poisons which they manufacture themselves. Sometimes the symptoms from these different causes can be similar and easily confused.

Apart from allergies, illness caused by eating molluscs is uncommon in Australia.

Allergies to Molluscs

Figure 71: Some people may develop allergies after eating molluscs such as these blue mussels *(Mytilus edulis planulatus)* from southern waters

A number of people learn from bitter experience that they are allergic to one or more types of molluscan foods, such as oysters, scallops, mussels, squid, etc. These unlucky people are affected after eating or even just touching foods which others can eat without any problem occurring. The basis of this different and often excessive reaction is similar to that involved in hay fever, and in some cases the symptoms may be similar.

A more detailed account of allergies appears in the section on Crustacea on page 65.

Allergy Prevention

Those types of molluscs which have caused problems in the past should be avoided.

If a person has developed an allergy to one type of mollusc, then other types should be tried with great caution, especially if they cause any swelling inside the mouth.

Allergic reactions are unpredictable and may develop for a whole group of related foods or for only one type. They may also change with time, sometimes becoming more severe but sometimes even improving.

Symptoms

- See under Crustaceans on page 65.
- Urticaria (swollen red and itchy skin).
- Swelling of part of the body including in the mouth and throat.
- Anaphylactic shock.

However, as with Crustaceans, some of these symptoms can result from poisoning or shock instead of from an allergy.

Treatment

- See under Crustaceans on page 65.
- Antihistamines by mouth or injection.
- Adrenalin for anaphylactic shock.
- Cortisone – locally as a cream
  – orally
  – by intravenous injection

Poisons from Pollution

Figure 72: Razor clams *(Pinna bicolor)* can concentrate pollutants as they filter water for food and oxygen.

Most farming and/or harvesting of molluscs such as oysters or scallops is carried out in shallow coastal waters, usually in sheltered embayments and estuaries where growth rates are highest. In such sheltered areas people also gather clams and cockles for eating. Unfortunately, these shallow coastal waters are sometimes polluted by domestic and industrial sewage, shipping harbour developments, and drainage from the adjacent countryside.

Molluscs from such areas, particularly filter-feeding bivalves, have caused many cases of illness. Attached to rocks or jetty piles and on or under the sand or mud, most bivalves feed by filtering particles of dead animal and plant tissue or living plankton from a water current which they draw through their perforated gills. Particles suitable for food are passed into the mouth while rejected material is tangled into mucus balls and flushed out of the bivalves. This feeding system can concentrate, though sometimes only temporarily, disease-causing micro-organisms, heavy metals and organic pollutants in the cavities or tissues of the bivalves. Some gastropods, such as periwinkles or creepers which feed upon encrusting or leafy algae or upon organic detritus, can also do this but less effectively.

Though most infective micro-organisms usually do not survive for long in the sea or inside molluscs, chemical pollutants can be further concentrated in carnivorous species such as whelks and some octopus and even humans which feed on molluscs from affected areas. In addition organic pollutants such as DDT, and inorganic pollutants such as zinc and other heavy metals can accumulate in such molluscs and be further concentrated in the human body sometimes to dangerous levels.

Injury Prevention

- Molluscs should not be eaten from areas likely to receive inadequately treated sewage. Molluscs from shallow areas frequented by seagulls should also be avoided, as these birds can carry disease-causing micro-organisms. Unfortunately molluscs are often both abundant and fast-growing in these areas, and so are often farmed or fished there. Holding suspect molluscs for 24 hours or more in clean salt water from a safe area, can reduce the danger from such micro-organisms. This is the basis of some commercial cleaning procedures.

Poisons Made by Molluscs or by the Food Organisms They Eat

Figure 73: The monitoring of PSP levels in oysters such as this large *Saccostrea echinata* will become necessary as oyster farming in northern Australian waters develops.

Very little is known of the chemical nature, origin or formation of the poisons which protect some, and perhaps many of the soft, fleshy and often brilliantly coloured nudibranchs, sea-hares and other gastropods in which the shell is thin and fragile or even absent. Some nudibranchs are known to use chemicals from the sponges they eat to produce distasteful and sometimes poisonous defensive secretions. At least some species of sea-hares are known to derive the poisons they produce from the leafy algae they eat. Such secretions from skin and other glands would probably serve a defensive function making the soft-bodied molluscs unpalatable to predators.

More is known about paralytic shellfish poison or PSP derived from substances produced by algae, and absorbed and concentrated by molluscs which are not affected themselves. One group of such toxins is produced by minute planktonic plants called dinoflagellates, particularly species of *Gonyaulax* and *Gymnodinium* found in temperate and tropical waters. Under certain climatic conditions the numbers of these dinoflagellates increase very quickly to form very high densities

Figure 74: Tropical reefs such as this at North West Cape, WA may be covered with tightly-packed coral rock oysters *(Saccostrea cucculata)*. This oyster species may retain PSP long after it has gone from other species nearby.

which colour the sea and make it fluoresce at night. Some give the water a red colour and such algal blooms are called red tides. At these times the level of toxins in bivalves feeding on these dinoflagellates becomes very high.

High levels of similar poisons occur, apparently spasmodically, in some weed-eating herbivorous gastropods such as turban and conch shells inhabiting tropical coral reefs, and in some carnivorous gastropods such as whelks and triton shells of more temperate waters. Dinoflagellate blooms cannot be the immediate origin of these toxins as these molluscs are not themselves filter feeders. The toxins are apparently mostly stored in the gastropods' digestive glands.

If these toxin-containing molluscs are eaten by humans or fish, symptoms of paralytic shellfish poisoning can develop, either rapidly or after a delay of a few hours. The severity depends on the amount of poison eaten and on the amount of other food in the stomach at the time. Different but often closely-related species of bivalves have different abilities to concentrate and store these toxins. A single oyster eaten from the Great Barrier Reef caused moderately severe paralytic shellfish poisoning, while another oyster of a different species, living less than 30 cm away, caused no ill effects when eaten at the same time.

Toxins of this group cause paralysis. One of them is a neurotoxin very like tetrodotoxin, the poison contained in puffer-fish. Its small molecules are not affected by the digestive process, and once absorbed into the blood stream the toxin affects the nervous system causing a loss of sensation and of controlled movement, along with general body pain. If skeletal muscle is sufficiently paralysed then breathing is affected and can even cease, causing death.

Poisoning Prevention

- The muscular parts of affected gastropods can, apparently, be eaten without ill effects if the visceral mass containing the digestive gland is cleanly removed. However, because the digestive gland is more widely spread through the tissues of bivalves, and because other organs can be used for toxin storage, no portion of affected bivalves such as oysters, cockles or giant clams should be eaten. These toxins are soluble in water and affected by heat so mincing, repeated washing and cooking (particularly as in canning) can reduce the toxin to tolerable concentrations.
- Unfortunately, neither the presence nor the concentration of paralytic shellfish toxin is indicated by any change in colour or odour of the affected molluscs.

Symptoms and Treatment

- The occurrence of paralytic shellfish poisoning is rare in Western Australian waters, although red tides have been reported off the Kimberley and Pilbara coasts. Should numbness, difficulty with speech and vision, weakness, increased salivation, and perhaps diarrhoea occur in one person after eating molluscs then vomiting should be induced in others who have eaten the same food. If paralysis develops, however, vomiting should not be induced as there is danger of vomit blocking the air passages. Mouth-to-mouth respiration should be used if respiratory distress is observed and should be continued until breathing is normal again or until a medical station is reached. The victim should never be left alone because of the danger of respiratory failure.

Echinoderms

Phylum Echinodermata

Figure 75: The crown-of-thorns starfish, *Acanthaster planci,* has razor sharp spines covered by thin skin containing venom glands.

Echinoderms are diverse in appearance ranging from stars to balls, but all have a spiny skin (hence the name which means just that), and an internal hard skeleton. This consists of loosely joined plates in starfishes, jointed arms in the feather stars and brittle stars, close-fitting plates in sea urchins and minute rods or ossicles embedded in the skin of sea cucumbers.

Most echinoderms are slow-moving, with a unique water-circulation system which operates rows of tube feet used for locomotion.

Some sea urchins and sea cucumbers are used for food by humans but many echinoderms contain toxic substances and others have venom glands associated with spines or special pincer organs called pedicellariae.

Feather Stars

Class Crinoidea

These are not known to be harmful to man but it is thought that some feather stars may secrete a toxic mucus.

Brittle Stars

Class Ophiuroidea

Figure 76: The only brittle star known to produce a toxic secretion is a common coral reef species, *Ophiomastix annulosa.*

Ophiomastix annulosa (Figure 76) is one of the few brittle stars known to be toxic. It produces a toxic mucus which causes paralysis and death in small animals, and so should be handled with care. It is common on tropical Indo-Pacific reefs including the atolls off the north-west coast.

Starfish

Class Asteroidea

Figure 77: Plectaster decanus may cause a rash, lasting for several days, if handled.

Many starfishes contain toxins in their digestive and reproductive organs which are not only poisonous when eaten but the poison can diffuse from living starfish into the water of an aquarium, killing molluscs, crustaceans and fish. Toxic water-soluble substances called saponins have been isolated from members of the families Asterinidae, Acanthasteridae and Asteriidae but may occur in other families that have not yet been tested. The saponins from various species had varying levels of toxicity when tested on fish in aquaria. All caused isolated human red blood cells to rupture.

Only two families of starfishes, Echinasteridae and Acanthasteridae, are known to have venomous spines. Echinasterids have a small body with long slender arms bearing spines on the net-like skeleton. Some species of *Echinaster* have thorny spines and surface pits from which a poisonous fluid is secreted. None of the six Australian species of *Echinaster* has caused injuries but *Plectaster decanus,* of the same family (Figure 77), found on the southern Australian coast from

New South Wales to Fremantle in Western Australia, has been recorded as causing contact dermatitis (Pope 1968).

Family Acanthasteridae

Figure 78: The underside of the crown-of-thorns starfish *(Acanthaster planci)*.

Crown-of-thorns Starfish *(Acanthaster planci)* (Figure 75)

The Crown-of-thorns starfish, *Acanthaster planci* (Figure 78), usually rare, has become notorious for its destruction of living coral over wide areas of tropical Indo-Pacific and Australian coral reefs when in plague numbers. It feeds on living coral polyps by spreading its stomach over the coral and digesting an area nearly as large as its body disc. It grows to a large size, commonly 30-40 cm across, rarely up to 60 cm across. The general colour of the starfish is usually blue-grey with a reddish tint when the respiratory papulae are expanded, while the spines are red to orange. It has a large disc with about 15-18 fairly short arms radiating from it. The disc and arms are covered with a soft skin and stout, hinged spines 2-3 cm long, each with a three-sided blade at the tip. The spines are covered by a thin skin containing two types of glands which are thought to produce venom and mucus.

Crown-of-thorns starfish sometimes lie in the open but are more often concealed under plate corals or ledges during the day, their colour blending with the background, so they may be unwittingly touched or trodden upon. In Western Australia the Crown-of-thorns starfish is common in the Dampier Archipelago, and rare on coral reefs as far south as Coral Bay, on the North West Cape peninsula.

The tissues contain toxic saponins, and are not only poisonous to humans but also to insects and soil organisms, suppressing plant growth. The starfish therefore cannot be used either for food or fertilizer.

Injury Prevention

- Wear gloves and protective clothing while diving or snorkelling on coral reefs. Wear boots if walking on a reef. Look first before putting your hand under a ledge.

Symptoms

- A puncture wound from a spine is intensely painful, causing swelling, redness, heat and numbness of the surrounding area.
- Stinging by 10 or more spines may result in vomiting which can recur every few hours for several days.
- If infection develops in the puncture wounds, lymph glands in the arm pit or groin may become tender or swollen.
- Often spine tips break off in the wound resulting in complications weeks or even months later. A spine tip in a finger can result in swelling and stiffness caused by the growth of granulation tissue typical of a foreign body reaction. The tissues may form a hard ball around the foreign body or bone-destroying (osteolytic) process may cause narrowing of a joint by the destruction of cartilage. Surgery is needed to correct both of these effects.

Treatment

- Any embedded pieces of spine should be removed to avoid later complications. Any suspicion that a piece of spine is actually in a joint represents a surgical emergency.
- Most marine poisons are destroyed by moderate heat. If possible immerse the affected part in hot water to 50°C, even water from a car radiator. The problems are heating the skin without scalding it, and getting the heat to the depth where the venom is.
- Thoroughly clean and disinfect the wound to prevent secondary infection. A bad stab warrants Tetanus toxoid to prevent the development of Tetanus. If protracted vomiting occurs try to stop it with anti-emetics (antihistamines) and attend to hydration.
- Medical attempts at pain-relief can include intravenous Pethidine, local injections of Lignocaine or a regional block anaesthesia (stunning the sensory nerve to the area) – to be done by an anaesthetist. Attempts at compression will not be very effective as they will tend to increase the pain.

Sea Urchins or Sea Eggs

Class Echinoidea

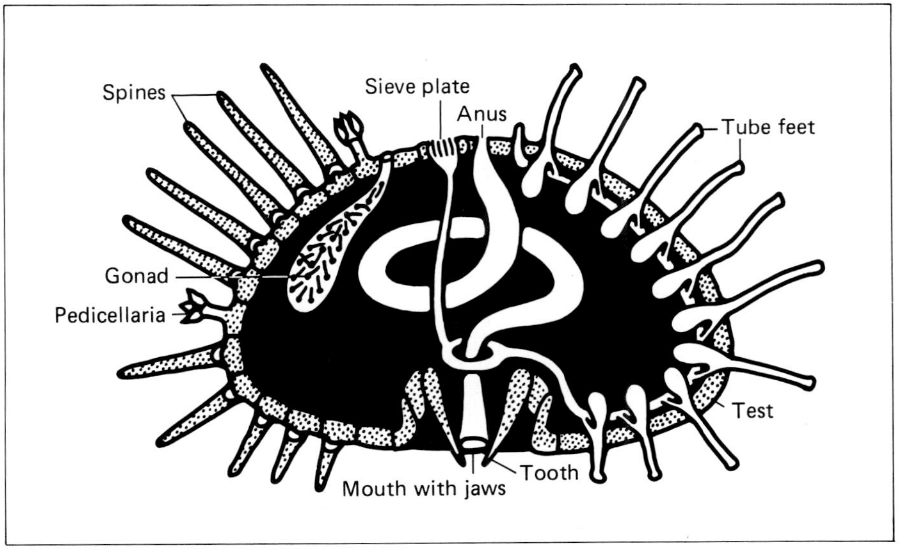

Figure 79: Diagrammatic section through a sea urchin showing spines on the left and tube feet on the right. In life the gut and five gonads occupy most of the body cavity.

These are circular or oval, spherical or flattened echinoderms with a shell of closely-fitting plates covered by a thin skin (Figure 79). Spines, from the thickness of a pencil to the thinness of a needle and from a few millimetres to more than 20 cm long project from the shell, articulating on rounded knobs which allow them to move for defence or locomotion. Like other echinoderms, hydraulically-operated tube feet with suckers at the tips are the main means of movement. The mouth on the lower surface has a complex jaw with five strong teeth which are used to crop the algae on which they feed. As well as spines, sea urchins have minute pincer organs or pedicellariae which remove foreign bodies from the skin. Most sea urchins can be handled safely but some have sharp venomous spines, and others have highly venomous pedicellariae (Figure 85).

SEA URCHINS WITH VENOMOUS LONG SPINES

Family Diadematidae

Figure 80: The needle-spined urchins, *Diadema* spp. can cause painful wounds with their spines which readily penetrate the skin and break off.

Diadema spp. (Figure 80)

Species of *Diadema* are tropical, needle-spined sea urchins found on coral reefs, rocky coasts and sandy bottoms in shallow water from low tide level to a depth of about 70 metres. Both *Diadema setosum* and *D. savignyi* are common in north-western Australia and as far south as the coral reefs of the Abrolhos off Geraldton. *D. savignyi* is occasionally found at Rottnest Island. They have a flattened circular shell (up to 10 cm wide) covered by black skin. The spines are hollow, brittle and very slender, reaching about 25 cm long. Shorter and thinner secondary spines are scattered among them. These urchins are usually black in the adult (occasionally with some white spines) but in juveniles the spines have black and white bands.

D. setosum has a reddish ring around the anus in the centre of the upper surface. Five blue spots and radiating white lines contrast with the black skin. The light-sensitive spots react to shadows or nearby movements, causing the spines to wave and converge on an intruder.

D. savignyi differs only by having a black anal area, a blue ring around the centre of the upper surface with five pairs of electric blue lines radiating from it, and the white spots may be absent.

Both species may group in large colonies during the day usually on sandy lagoon bottoms, or under rocks, and disperse at night to feed on algae.

No toxin has been isolated from the spines but the reaction of humans indicates that there is a venom, probably in the thin skin covering the spines. The long needle-spines are very fragile and very sharp, and readily penetrate the skin and break off. Their rough surface and brittle texture make them almost impossible to remove (Figure 81a).

Symptoms

- Severe pain, lasting from half an hour to four hours. Swelling and inflammation may occur.
- Fragments of spines may be visible under the skin.
- The injury may be complicated by a foreign body reaction to small detached pieces of spine or the skin covering the spine.

Treatment

There are several conflicting ideas.

- Western practice demands the removal of the foreign body. This is difficult because of its extreme brittleness, so that a spine can rarely be removed intact. A thorough-going surgical approach may include x-rays because the spines are composed of calcium carbonate and show up well. However, spine fragments of *Diadema* will usually be absorbed without harm, or work their way out.
- A traditional Indonesian treatment is to pummel the area with a smooth stone which breaks up the spines into small pieces so that they may be absorbed. The pieces would need to be very small.
- A competing idea is to soak the skin with vinegar and leave it open and hope that there will be pustule formation, and that the pieces will come out in the pustules.
- During the time of intense pain, heating the area with hot water may inactivate the supposed venom.

Take measures for pain-relief and treatment of secondary infection.

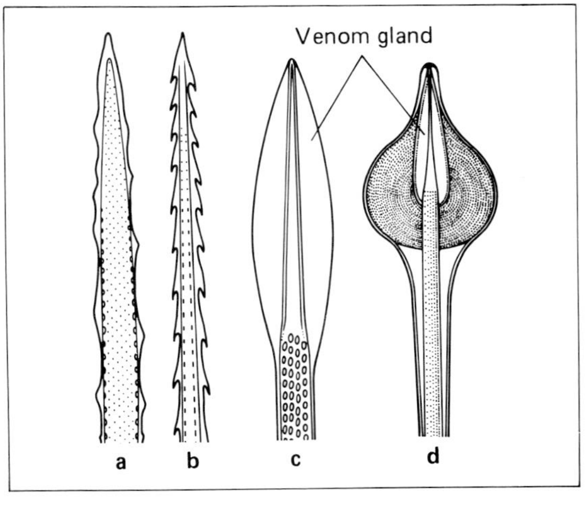

Figure 81: Tips of spines of (a) *Diadema setosum* (x33), (b) *Echinothrix diadema* (x33), (c) *Araeosoma thetidis*, enclosed in a venom gland (x27), (d) *Asthenosoma varium*, with venom gland (x27). (After Mortensen 1935, 1940).

Echinothrix spp.

Species of *Echinothrix* are readily distinguished from *Diadema* by five bands of very fine, golden-brown to black secondary spines extending from the upper to the lower surface of the body. The primary spines are shorter than those of *Diadema* and in adults are less than half as long as the shell width which may be up to 15 cm. In small examples the spines are often longer than the width of the shell.

Figure 82: Echinothrix spp. have shorter primary spines than *Diadema* and the very fine, barbed, venomous secondary spines are grouped in bands.

Echinothrix calamaris (Figure 82), particularly a young specimen, can be a beautiful sea urchin with black, white or green, often banded, primary spines and very fine, golden-brown secondary spines. Very large specimens are often entirely black. Small white plates are embedded in the skin around the anus.

The primary spines are hollow with a large cavity, making them extremely brittle. They are covered by whorls of minute scales, the free ends of which point towards the spine tip. The secondary spines have barbs pointing away from their tips (Figure 81b) which make them impossible to pull out if they become embedded in a victim. The tips are encased in skin containing venom glands.

Echinothrix diadema specimens are less brightly coloured than those of *E. calamaris* and small specimens often have dark green primary spines banded with black, while the spines of large specimens are usually dark coloured. The hollow primary spines have only a small cavity, making them less brittle than those of *E. calamaris,* and the scales are on ridges running along the spines and not in whorls around them. There are no white plates near the anus.

Both species live in shallow water on coral reefs, often partly concealed under boulders or ledges. They are both found throughout the tropical Indo-West Pacific and on the northern, eastern and western coasts of Australia southwards to the Abrolhos in Western Australia.

Venom glands at the tips of the secondary spines of both species contain noradrenaline which affects blood pressure, and a protein which is believed to cause the intense pain.

Symptoms

- Immediate severe burning pain which becomes throbbing as redness and swelling develop.
- Fragments of spines may be visible particularly as the surrounding skin can become blue stained.
- Secondary infection of the puncture wounds is common.
- Delayed reactions from embedded spine fragments may occur.

Case 1

- In a Western Australian case a spine in the fingertip caused intense pain and numbness which spread to the whole hand within minutes with throbbing pain running up the arm. After half an hour the whole arm was numb but after an hour the numbness decreased, persisting in the finger for 24 hours. The area around the entry point of the spine turned bright blue initially, the colour spreading to the whole finger, fading to purple and retreating to the fingertip after half an hour.

Case 2

- Fragments of spines in a toe caused pain and swelling of the foot three months after penetration and the pieces of spine had to be removed under general anaesthetic. The spines are not absorbed like those of *Diadema* and unremoved fragments may be seen by x-ray many months after the original injury. This injury, described by Earle (1941), was erroneously attributed to *Diadema.* His illustration shows a species of *Echinothrix* and confusion between the symptoms of *Diadema* and *Echinothrix* injuries have persisted in the literature.

Treatment

- Since the spines are venomous and are not absorbed the treatment differs from that of *Diadema* injuries.
- Prompt surgical removal of spines is desirable. Prior to the operation the injured part should be handled as little as possible to avoid fragmentation of the spines (quite unlike treatment for *Diadema* spines).
- Bathing the affected area in water as hot as can be borne may ease the pain.
- Apply antiseptic and a local anaesthetic ointment to the wound.

Figure 83: The common needle-spined urchin in south-western Australia is *Centrostephanus tenuispinus*. Although it has sharp brittle spines and venomous pedicellariae no injury from this species has been recorded.

Centrostephanus tenuispinus (Figure 83)

This needle-spined urchin is common on the south-west coast of Australia, from Shark Bay to South Australia. It is not found elsewhere, although a related species lives on the east coast of Australia. It is a large sea urchin (up to 11 cm across) and has hollow, scaly spines 7 cm long. The shell is covered with a dark skin of reddish-brown to dull greenish or purple, and the spines are lavender grey to nearly black.

Centrostephanus lives among rocks or under ledges on coral or limestone reefs subtidally to 100 metres. Unlike other diadematid urchins it has pedicellariae with venom glands. No case of stinging by this species has been reported but it should be handled with care both because of its needle-spines and its venomous pedicellaries.

Treatment

- As for *Diadema.*

SEA URCHINS WITH VENOMOUS SHORT SPINES

Family Echinothuriidae

Figure 84: Asthenosoma spp. have short, fine, very sharp spines encased in venom glands. They sting severely at the slightest touch.

Tam-o'shanter Urchins (*Asthenosoma* and *Araeosoma* species)(Figure 84).

These are large (up to 14 cm across), brightly coloured, short-spined sea urchins which have a thin shell which collapses out of water giving them a 'beret'-like appearance (hence the common name). All have highly venomous spines which are fairly short, thin, extremely sharp and able to penetrate most types of gloves (Figure 81c, d).

In *Asthenosoma* the spines of the upper side are covered by an inflated skin sheath, sometimes white but often pale coloured at the base with purplish bands shading to pale to brilliant electric-blue at the tips. The spines each have a single large poison sac at the tip and venom is injected when the spines penetrate the skin and, usually, break off in the wound. Species of *Asthenosoma* occur from low tide level to about 150 metres. Two Indo-Pacific species, *A. varium* and *A. ijimai,* are found in northern Australia, southwards to the coral reefs of the Houtman Abrolhos.

Two species of *Araeosoma, A. owstoni* and *A. tesselatum,* have been trawled from 150-400 metres depth on the North-West Shelf. They are similar in appearance to species of *Asthenosoma.*

The composition of the venom is not known.

Injury Prevention

- Do not pick up these sea urchins even with heavy gloves.

Symptoms

- Immediate intense pain with swelling, lasting several hours.

Treatment

- Immerse the affected part in water as hot as can be borne.
- Surgical removal of spine fragments is desirable.
- Local anaesthetic.

SEA URCHINS WITH VENOMOUS PEDICELLARIAE

Family Toxopneustidae

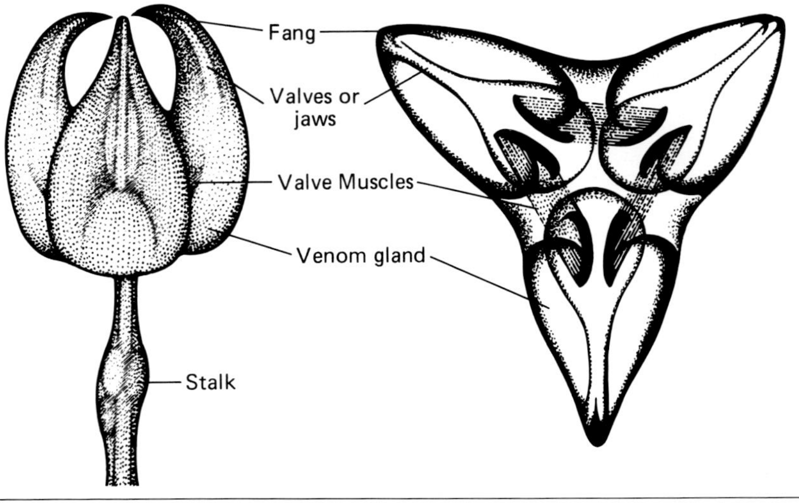

Figure 85: A venomous pedicellaria of *Toxopneustes* showing the needle-sharp fangs connected to venom glands in the three jaws (x15) (after Mortensen, 1943).

The pincer organs or pedicellariae of sea urchins are minute, often pin-head sized, seizing organs scattered among the spines. Some are used for defence while other types clean the body surface. They consist of three hard jaws on a flexible stalk. Pedicellariae occur in various shapes and sizes, characteristic of each species, but one kind, the globiferous pedicellariae (Figure 85) has venom glands and sharp, fang-like teeth at the ends of the jaws. Many sea urchins have globiferous pedicellariae for defence but they are usually too small to be harmful to humans, however one species has very large, highly venomous pedicellariae and is dangerous. Several other species are less toxic but should be handled with care.

Figure 86: The short spines of *Toxopneustes pileolus* are hidden by the open flower-like, highly venomous pedicellariae.

Toxopneustes pileolus (Figure 86)

This tropical sea urchin lives on coral reefs and sand or mud bottoms from the intertidal zone to a depth of 90 metres. It is a widespread Indo-Pacific species found in northern Australia and as far south as Moreton Bay in Queensland and Shark Bay in Western Australia.

The body is a low round dome, with a concave underside. It can be up to 15 cm wide, but 6-10 cm is more usual. The spines of the upper side are 6-8 mm long, and those on the lower side 10-12 mm long. The spines are fairly robust, taper to a point, and are in shades of red, green or white, or sometimes bands of all three colours.

In life, the spines are almost concealed by the open, flower-like globiferous pedicellariae, each of which has a rounded triangular shape and is up to 3 mm across (Figures 85 and 86). When sensory hairs on the inner side of the jaws of the pedicellariae are stimulated, the jaws close and inject venom through the fangs. They cannot bite through the thick skin of the palm of the hand but the side of a finger or tender skin is susceptible. Bites by several pedicellariae cause severe pain and general symptoms. Drowning, as a result of being bitten, has been reported in Japan.

The venom interferes with nerve-muscle conduction and paralyses the muscles. It also increases capillary permeability – which can be reversed by antihistamine drugs.

Injury Prevention

- Wear gloves if you need to handle *Toxopneustes.*

Symptoms

- Pain, severe and spreading, lasting up to an hour.
- Collapse, paralysis and breathing difficulty.

Case Report

- A bite by seven or eight pedicellariae on the side of a finger caused severe pain followed by giddiness, difficult breathing, paralysis of the lips, tongue and eyelids, and relaxation of the muscles in the limbs. After 15 minutes the pain gradually diminished and disappeared after an hour. The face was paralysed for about six hours (Fujiwara 1935, quoted by Mortensen 1943).

Treatment

- This is directed to relief of pain and the immediate first aid is to bathe the affected part in water as hot as the patient can stand.
- Regional block anaesthesia or intravenous Pethidine may be used for pain relief.
- If respiratory distress occurs with the skin becoming bluish in colour, carry out mouth-to-mouth artificial respiration. If artificial respiration does not correct the patient's colour or if no pulse or heartbeat can be felt, apply external cardiac massage.
- Treat for shock.
- Seek medical assistance. Antihistamine drugs reverse some of the effects.

Figure 87: Tripneustes gratilla, also with venomous pedicellariae, is common on coral reefs but this specimen, flanked by colonies of *Palythoa* was photographed at Rottnest Island, near Fremantle.

Tripneustes gratilla (Figure 87)

This very common tropical Indo-West Pacific sea urchin lives on coral reefs and limestone platforms as far south as Rottnest Island in Western Australia. It has a depth range from the intertidal zone to about 70 metres. *T. gratilla* is usually concealed under ledges, or covers itself with fragments of algae and shells. The cleaned shell is often pale lavender in colour but when alive short white spines stand out conspicuously against the dark skin covering the shell. In other parts of the Indo-Pacific the spines may be orange or even black. There are many very small globiferous pedicellariae amongst the spines but with normal handling these are too small to bite. However, if one becomes attached to the tender skin of a finger or inside of the arm it may sting quite severely. The gonads of *Tripneustes gratilla* are edible but should be well washed to remove any pedicellariae which may adhere to them.

The venom from the pedicellariae contains a water soluble protein which lowers the blood pressure and ruptures red blood cells. It also contains a chemical which affects the heart, and an enzyme which catalyzes histamine release. It is precipitated by 65% saturated ammonium sulphate and is deactivated by heating to between 45° and 47.5°C, like most marine poisons.

Injury Prevention
- Wear gloves when handling *Tripneustes.*
- Wash the gonads thoroughly before eating them.

Symptoms
- Immediate pain, localised swelling and redness, and an aching sensation in the affected part.

Cases:
- A sting on the tender skin inside the elbow by pedicellariae on a fragment of shell was more painful than nettles and lasted several hours (Mortensen 1943). The resulting wound, similar to a burn, did not heal for more than a month.
- The pain from a sting by a single pedicellaria was similar to a bee sting. A minute swelling at the site of each of the three puncture marks was followed by a weal 1 cm across. The symptoms lasted about one hour and there were no general effects. Subsequent stingings reacted more severely, lasting up to eight hours, with a weal 12 cm across indicating an increasing histamine response (Alender 1966).
- Intense pain and swelling of the lips and mouth has been experienced after eating gonads which have not been thoroughly washed. This venom probably came from pedicellariae which attached themselves to the gonad as it was being removed.

Treatment
- Bathe the affected part in water as hot as the patient can stand.
- Treat with antihistamine.
- Apply 20% aluminium sulphate, if it can get to where the venom is, it may help.

Other sea urchins with venomous globiferous pedicellariae that have been reported to sting humans are:

Heterocentrotus mammillatus

This slate-pencil sea urchin, found on coral reefs, is only known at present in Western Australia from atolls far off the north-west coast, but it is moderately common on the Great Barrier Reef off Queensland.

Salmacis sphaeroides

This sea urchin is fairly common on shallower parts of the North-West Shelf off Western Australia and is often taken in fish and prawn trawls. It has fairly large globiferous pedicellariae, but stings have been reported from the spines.

Sea Cucumbers

Class Holothuroidea

Figure 88: Actinopyga mauritiana is one of many large sea cucumbers known as beche-de-mer or trepang when prepared for eating. However all the edible species are toxic unless correctly prepared.

Holothurians, Sea Cucumbers, Beche-de-mer, Trepang

These are sausage-shaped animals, from a few centimetres to over a metre in length (Figure 88). Some lie inert on sandy bottoms using short sticky tentacles to collect food particles from the sand, while others spread long branching tentacles in the water to trap plankton. Like other echinoderms they move about on tube feet which extend through the body wall. They are most abundant in tropical seas where a number of shallow water species are collected and processed for food, used by Chinese, Japanese and the Pacific Islanders. The large species used for food have thick body walls and belong to the families Holothuriidae and Stichopodidae. The prepared product is known as trepang (the Malay name) or beche-de-mer.

Over 40 species of sea cucumber that have been tested, including all those used for food, contain toxic triterpenoid saponins of several types which irreversibly block nervous and muscular action. The poisons dissolve in water but are not destroyed by heat. Dried sea cucumber products have little toxicity (Hashimoto 1977), as lengthy contact with water during processing leaches most of the poison from the body wall. Processing includes gutting, boiling for up to 1.5 hours, washing and drying.

Many species of sea cucumber, when disturbed, rapidly eject sticky white threads from the anus. These Cuvierian tubules are highly toxic and those of some species, for example the large leopard or tiger fish *Bohadschia argus*, cause inflammation if they touch the skin.

Holothurians are most common on the tropical reefs of Western Australia but some species occur along the west and south coasts.

Injury Prevention

- Beche-de-mer should not be eaten unless they have been correctly prepared.
- Avoid touching the white threads ejected from the anus and if these are touched be careful not to touch the face, especially the eyes.

Sea Squirts or Ascidians

Phylum Chordata
Subphylum Tunicata
Class Ascidiacea

Figure 89: The large pink sea-squirt, *Herdmania momus,* has the inner body wall strengthened with fine calcareous spicules which can penetrate the hands.

Ascidians are quite complex animals, deceptively simple in appearance. They are sac-like, solitary or colonial animals which live attached to rocks or jetty piles and are sometimes used for fishing bait. The common name arises from their ability to contract the muscular inner body wall squirting a jet of water through one or both of the external openings. Water is taken in through an inhalent siphon to a basket-like pharynx where minute food particles are sieved out and passed on to the stomach. Oxygen is extracted as water passes through the pharynx into a surrounding chamber which leads to an exhalent siphon from which it is expelled.

The soft internal organs are enclosed in a thin but muscular body wall which is protected by an outer 'tunic' composed of a cellulose-like substance, which in some species is thick and tough. In the large pink sea squirt, *Herdmania momus,* which may be 10-15 cm high and as much across, the inner body wall is strengthened with a mat of calcareous needle-like spicules which can painfully penetrate the hands if the animal is cut or damaged (Figures 89, 90).

Ascidians lack a backbone but have tadpole-like larvae with an elastic skeletal rod and gill clefts similar to those possessed by larval vertebrate animals. For these reasons they are included in the phylum Chordata, to which fishes and man belong.

Injury Prevention

- Wear gloves if handling sea squirts.

Treatment

- The spicules, composed of calcium carbonate, dissolve rapidly in weak acid therefore soaking the affected part in vinegar for a few minutes should dissolve them and relieve the pain.

Or apply adhesive tape to the affected area and pull of gently with the attached spicules.

Or pull them out with tweezers with the help of a magnifying glass.

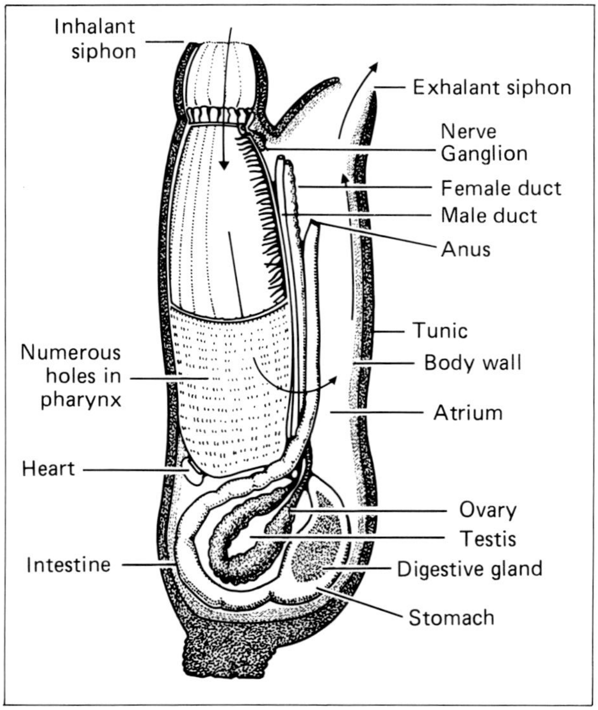

Figure 90: A sea-squirt with the tunic and body wall removed from one side to show the internal organs. Water passes through the holes in the pharynx to the atrium and out by the exhalent siphon while food is collected on one side of the pharynx and passed down to the stomach. The anus and reproductive ducts all open into the atrium.

Further Reading

General

Bennett, I. (1966). *The fringe of the sea.* Rigby Limited: Adelaide.

Catala, R. (1964). *Carnival under the sea.* R. Sicard: Paris.

Cleland, J.B. and Southcott, R.V.(1965). *Injuries to man from marine invertebrates in the Australian region.* National Health and Medical Research Council, Special Report Series No. 12. Canberra.

Edmonds, C. (1975). *Dangerous marine animals of the Indo-Pacific Region.* Wedneil Publications: Newport, Victoria.

Edmonds, C. (1984). *Marine animal injuries to man.* Wedneil Publications: Newport, Victoria.

Halstead, B.W. (1965). *Poisonous and venomous marine animals of the world* **1**: U.S. Govt Printing Office: Washington, D.C.

Hashimoto, Y. (1977). *Marine toxins and other bioactive marine metabolites.* Japan Scientific Societies Press: Tokyo. [English edition, 1979].

Henderson, D. and Easton, R.G. (1980). Stingose, a new and effective treatment for bites and stings. *Med. J. Aust.* 1980 **2**: 146-150.

Keegan, H.L. and MacFarlane, W.V. (eds)(1963). *Venomous and poisonous animals and noxious plants of the Pacific area.* Pergamon Press: Oxford, London, New York.

Pearn, J. (ed.)(1981). *Animal toxins and man.* Div. of Health Education and Information. Queensland Health Dept: Brisbane.

Russell, F.E. (1984). Marine toxins and venomous and poisonous marine plants and animals (Invertebrates). *Adv. Mar. Biol.* **21**: 59-230.

Sutherland, S. (1980). *Family guide to dangerous animals and plants of Australia.* Oxford: Adelaide.

Sutherland, S. (1982). *Venomous creatures of Australia: a field guide with notes on first aid.* Oxford: Melbourne.

Sutherland, S. (1983). *Australian animal toxins: the creatures, their toxins and care of the poisoned patient.* Oxford Univ. Press: Melbourne.

Sponges

Bergquist, P. (1982). Porifera. In *Marine invertebrates of southern Australia.* (S.A. Shepherd and I.M. Thomas, eds). South Australian Govt Printer: Adelaide.

Southcott, R.V. and Coulter, J.R. (1971). The effects of the southern Australian marine stinging sponges, *Neofibularia mordens* and *Lissodendoryx* sp. *Med. J. Aust.* 1971 **2**: 895-901.

Cnidarians (Coelenterates)

Barnes, J.H. (1966). Studies on three venomous cubomedusae. In *The Cnidaria and their evolution. Symp. zool. Soc. London* No. 16. (W.J. Rees, ed.): 307-332.

Burnett, J.W. and Calton, G.J. (1974). Sea nettle and man-o'-war venom: chemical comparison of their venoms and studies on the pathogenesis of the sting. *J. Invest. Dermatol.* **62**: 372-377.

Burnett, J.W. and Calton, G.J. (1977). The chemistry and toxicology of some venomous pelagic coelenterates. *Toxicon* **15**: 177-196.

Dunn, D.F. (1981). The clownfish sea anemones: Stichodactylidae (Coelenterata, Actiniaria) and other sea anemones symbiotic with pomacentrid fishes. *Trans. Amer. phil. Soc.* **71** (1): 1-115.

Endean, R. (1981). The box-jellyfish or 'sea wasp'. In *Animal toxins and man.*
Pearn, J. (ed.)(1981). *Animal toxins and man.* Div. of Health Education and In-

formation. Queensland Health Dept: Brisbane.

Endean, R., Duchemin, C., McColm, D. and Fraser, E.H. (1969). A study of the biological activity of toxic material derived from nematocysts of the cubomedusan *Chironex fleckeri. Toxicon* **6**: 179-204.

Freeman, S.E. and Turner, R.J. (1972). Cardiovascular effects of cnidarian toxins: a comparison of toxins extracted from *Chiropsalmus quadrigatus* and *Chironex fleckeri. Toxicon* **10**: 31-37.

Hartwick, R., Callanan, V. and Williamson, J. (1980). Disarming the box-jellyfish – nematocyst inhibition in *Chironex fleckeri. Med. J. Aust.* 1980 **1**: 15-20.

Mebs, D. and Gebauer, E. (1980). Isolation of proteinase inhibitory, toxic and hemolytic polypeptides from a sea anemone *Stoichactis* sp. *Toxicon* **18**: 97-106.

Neeman, I., Calton, G.J. and Burnett, J.W. (1980). Cytotoxicity and dermonecrosis of sea nettle *(Chrysaora quinquecirrha)* venom. *Toxicon* **18**: 55-63.

Shryock, J.C. and Bianchi, C.B. (1983). Sea nettle *(Chrysaora quinquecirrha)* nematocyst venom; mechanism of action on muscle. *Toxicon* **21** (1): 81-95.

Southcott, R.V. (1956). Studies on Australian Cubomedusae, including a new genus and species apparently harmful to man. *Aust. J. Mar. Freshw. Res.* **7**: 254-280.

Southcott, R.V. (1967). Revision of some Carybdeidae (Scyphozoa: Cubomedusae), including a description of the jellyfish responsible for the 'Irukandji syndrome'. *Aust. J. Zool.,* **15**: 651-71.

Southcott, R.V. (1968). The Portuguese man-of-war or bluebottle. *The Australian Museum, Sydney,* Leaflet **68**: 1-11.

Warwick, J.E., Weinreich, D. and Burnett, J.W. (1981). Sea nettle *(Chrysaora quinquecirrha)* toxin on electrogenic and chemosensitive properties of nerve and muscle. *Toxicon* **19** (3): 361-371.

Williamson, J.A., Callanan, V.I. and Hartwick, R.F. (1980). Serious envenomation by the northern Australian box-jellyfish *(Chironex fleckeri). Med. J. Aust.* 1980 **1**: 13-15.

Swimmers Itch

Appleton, C.C. (1980). Studies on the ecology of *Austrobilharzia terrigalensis* (Trematoda: Schistosomatidae) in the Swan Estuary. Western Australia. Ph.D. Thesis, Murdoch University (unpublished).

Appleton, C.C. and Lethbridge, R.C. (1979). Schistosome dermatitis in the Swan Estuary, Western Australia. *Med. J. Aust.* 1971 **1**: 141-144.

Bearup, A.J. (1956). Life cycle of *Austrobilharzia terrigalensis* Johnston, 1917. *Parasitology* **46**: 470-479.

Pope, E.C. (1972). Terrigal itch, a coastal waters dermatitis. *Aust. Nat. Hist.* **17** (7). 217-221.

Crustacea

Raj, U., Haq, H., Oshima, Y. and Yasumoto, T. (1983). The occurrence of paralytic shellfish toxins in two species of xanthid crab from Suva Barrier Reef, Fiji Islands. *Toxicon* **21** (4): 547-551.

Molluscs

Altman, J.S. and Nixon, M. (1970). Use of the beaks and radula by *Octopus vulgaris* in feeding. *J. Zool.* **161**: 25-38.

Bidder, A.M. (1966). Feeding and digestion in cephalopods. In *Physiology of Mollusca* 2 (K.M. Wilbur and C.M. Yonge, eds): 97-124. Academic Press: New York.

Boyle, P.R. and Knobloch, D. (1981). Hole boring of crustacean prey by the octopus *Eledone cirrhosa* (Mollusca, Cephalopoda), *J. Zool.* **193** (1): 1-10.

Burrenson, B.J., Scheuer, P.J., Finer, J. and Clardy, J. (1975). 9-Isocyano-pupukeanane, a marine invertebrate allomone with a new sesquiterpene skeleton, *J. Amer. Chem. Soc.* **97**: 4763-4764. *(Phyllidia varicosa* concentrates terpenoid compounds from sponges.)

Croft, J.A. and Howden, M.E.H. (1972). Chemistry of maculotoxin: a potent neurotoxin isolated from *Hapalochlaena maculosa, Toxicon* **10**: 645-651.

Cummings, B. (1936). Encounter between cone shells and octopus, *North Queensl. Nat.* **4**: 42.

Endean, R. (1964). Venomous cones. *Aust. nat. Hist.* **14** (12): 400-403.

Endean, R. (1981). The venom of the blue-ringed octopus. In *Animal toxins and man.* (J. Pearn, ed.). Division of Health Education and Information, Queensland Health Department: Brisbane.

Endean, R. and Duchemin, C. (1967). The venom apparatus of *Conus magus. Toxicon* **4**: 275-284.

Endean, R., Gyr, P. and Parish, G. (1974). Pharmacology of the venom of the gastropod *Conus magus. Toxicon* **12**: 117-129.

Endean, R., Gyr, P. and Surridge, J. (1979). The effects of crude venom of *Conus magus* and *C. striatus* on the contractile response and electrical activity of guinea pig cardiac musculature. *Toxicon* **17** (4): 381-395.

Endean, R., Izatt, J. and McColm, D. (1967). The venom of the piscivorous gastropod *Conus striatus.* In *Animal toxins:* 137-144. Pergamon Press: Oxford and New York.

Endean, R., Parish, G. and Gyr, P. (1974). Pharmacology of the venom of *Conus geographus. Toxicon* **12** (2): 131-138.

Endean, R. and Rudkin, C. (1963). Studies on the venoms of some Conidae. *Toxicon* **1**: 49-64.

Endean, R. and Rudkin, C. (1965). Further studies on the venoms of Conidae. *Toxicon* **2**: 225-249.

Endean, R., Williams, H., Gyr, P. and Surridge, J. (1976). Some effects on muscle and nerve of crude venom from the gastropod *Conus striatus. Toxicon* **14**: 267-274.

Flecker, H. and Cotton, B.C. (1955). Fatal bite from octopus. *Med. J. Aust.* 1955 **2**: 329-331.

Harris, J. (1981). Tetrodotoxin. In *Animal toxins and man.* (J. Pearn, ed.): 25-28. Division of Health Education and Information, Queensland Health Department: Brisbane.

Hermitte, L.C.D. (1946). Venomous marine molluscs of the genus *Conus. Trans. Roy. Soc. Trop. Med. Hyg.* **39**: 485-512, 1, 3, 4, 5 pls.

Hinegardner, R.T. (1958). The venom apparatus of the cone shell. *Hawaii Med. Jour.* **17**: 533-536.

Kohn, A.J. (1956). Piscivorous gastropods of the genus *Conus. Proc. Natl Acad. Sci. U.S.* **42**: 168-171.

Kohn, A.J. (1959). The ecology of *Conus* in Hawaii. *Ecol. Monogr.* **29**: 47-90.

Kohn, A.J. (1963). Venomous marine snails of the genus *Conus.* In *Venomous and poisonous animals and noxious plants of the Pacific area.* (H.L. Keegan and W.V. MacFarlane, eds). Pergamon Press: Oxford, London, New York, Paris.

Kohn, A.J., Saunders, P.R. and Wiener, S. (1960). Preliminary studies on the venom of the marine snail *Conus. Ann. N. York Acad. Sci.* **90** (3): 706-725.

Kotari, Y., Oshima, Y. and Yasumoto, T. (1981). Analysis of paralytic shell-fish toxins of marine snails. *Bull. Jap. Soc. Scient. Fish.* **47** (7): 943-946.

McMichael, D.F. (1971). Mollusks – classification, distribution, venom apparatus and venoms, symptomatology of stings. In *Venomous animals and their venoms.* **3.** (W. Bucherl and E.E. Buckley, eds): 373-393. Academic Press: New York, London.

Marsh, H. (1970). Preliminary studies on the venom of some vermivorous Conidae. *Toxicon* **8**: 271-277.

Narita, H., Noguchi, T., Maruyama, J., Eeda, Y., Hashimoto, Y., Watanabe, Y. and Hida, K. (1981). Occurrence of tetrodotoxin in a trumpet shell, 'Boshubora' *Charonia sauliae. Bull. Jap. Soc. Scient. Fish.* **47** (7): 935-941.

Nixon, M. and Messenger, J.B. (eds)(1977). *The biology of cephalopods. Symp. Zool. Soc. Lond.* No. 38. Academic Press: London, New York and San Francisco.

Noguchi, T., Maruyama, J., Ueda, Y., Hashimoto, K. and Harada, T. (1981). Occurrence of tetrodotoxin in the Japanese ivory shell *Babylonia japonica. Bull. Jap. Soc. Scient. Fish.* **47** (7): 909-913.

Overath, H. and Boletzky, S.V. (1974). Laboratory observations on spawning and embryonic development of a blue-ringed octopus. *Mar. Biol.* **27**: 333-337.

Robson, G.C. (1929). *A monograph of the Recent Cephalopoda. Pt 1. Octopodinae.* Trustees of the British Museum: London. 1-235.

Savage, I.V.E. and Howden, M.E.H. (1977). Hapalotoxin, a second lethal toxin from the octopus *Hapalochlaena maculosa. Toxicon* **15** (5): 463-466.

Scrymgeour, A. (1985). A cone in my bedroom. *Underwater* **12**: 57-58.

Sheumack, D.D., Howden, M.E.H., Spence, I. and Quinn, R.J. (1978). Maculotoxin: a neurotoxin from the venom glands of the octopus *Hapalochlaena maculosa* identified as tetrodotoxin. *Science* **199**: 188-189.

Sutherland, S.K. and Lane, W.R. (1969). Toxins and mode of envenomation of the common ringed or blue banded octopus. *Med. J. Aust.* 1969 **1**: 893-898, 7 figs and 1 pl. in suppl.

Sutherland, S.K., Broad, A.J. and Lane, W.R. (1970). Octopus neurotoxins: low molecular weight non-immunogenic toxins present in the saliva of the blue-ringed octopus. *Toxicon* **8**: 249-250.

Tranter, D.J. and Augustine, O. (1973). Observations on the life history of the blue-ringed octopus *Hapalochlaena maculosa. Mar. Biol.* **18**: 115-128.

Wells, M.J. (1978). *Octopus: physiology and behaviour of an advanced invertebrate.* Chapman and Hall, London: i-xiii, 1-417.

Wells, M.J., Freeman, N.H. and Ashburner, M. (1965). Some experiments on the chemotactile sense of octopuses. *J. exp. Biol.* **43**: 553-563.

Whyte, J.M. and Endean, R. (1962). Pharmacological investigation of the venoms of the marine snails *Conus textile* and *Conus geographus. Toxicon* **1**: 25-31.

Echinoderms

Alender. C.B. and Russell, F.E. (1966). Pharmacology. In *Physiology of Echinodermata,* (R.A. Boolootian, ed.): 529-543. Interscience: New York.

Barnes, J. and Endean, R. (1964). A dangerous starfish – *Acanthaster planci* (Linnaeus). *Med. J. Aust.* 1964 **1**: 592-593.

Earle, K.V. (1941). Echinoderm injuries in Nauru. *Med. J. Aust.* 1941 **2**: 265-266.

Kimura, A. and Nakagawa, H. (1980). Action of an extract from the sea urchin, *Toxopneustes pileolus,* on isolated smooth muscle. *Toxicon* **18**: 689-693.

Mebs, D. (1984). A toxin from the sea urchin *Tripneustes gratilla. Toxicon* **22** (2): 306-307.

Mortensen, T. (1935-1943). *A monograph of the Echinoidea.* Vols II, III. C.A. Reitzel: Copenhagen.

Pope, E.C. (1964). A stinging by a crown-of-thorns starfish. *Aust. nat. Hist.* **14**: 350.

Pope, E.C. (1968). Venomous starfish in Sydney Harbour. *Aust. Nat. Hist.* **16** (1): 26.

Vasserot, J. (1980). Signification écologique de la fonction vénimeux chez les Echinodermes. In *Echinoderms present and past* (M. Jangoux, ed.): 343-347. A.A. Balkema, Rotterdam.

First Aid

Allergies

Symptoms can range through hay fever (sneezing, running eyes and nose), urticaria (swelling of large areas of red itchy skin), angio-neurotic oedema (pale swelling which can be serious if in the throat) to anaphylactic shock (obviously serious, with collapse and breathing difficulty). These can occur in some people after eating or even touching

- crustaceans – such as prawns, crabs, rock lobsters etc. See pages 64 to 66.
- molluscs – such as mussels, whelks etc. See pages 96, 97 and 65, 66.
- Portuguese man-of-war. See pages 15 and 16.

First Aid

1. Take antihistamine (hay fever or sea sickness tablets).
2. Apply hydrocortisone creams.
3. For more severe cases seek medical treatment.

Bites Causing Inflammation

Pain swelling and bleeding may be caused by

- some cone shells. See pages 85 and 86.
- squid, cuttlefish and some octopus. See pages 89 and 90.

First Aid

1. Clean and close wound.
2. Give Tetanus booster.

Bites Causing Paralysis

Effects on nerves and muscles can range from local numbness to difficulty in swallowing and speaking and visual problems to overall paralysis of muscles including those for breathing. Symptoms can occur after the bite (sometimes unnoticed) of

- some cone shells. See pages 86 and 87.
- blue-ringed octopus. See pages 94 and 95.

First Aid

1. Seek medical help *but do not leave victim alone.*
2. Apply pressure bandage over area of bite and beyond.
3. Immobilise limb and/or body.

4. Reassure victim. Explain. Talk even though he/she cannot reply or respond.
5. Use artificial respiration if breathing fades.
6. Use cardiac massage if heart fades.
7. Place victim in Coma Position, lying on side. Watch for vomiting and choking, clear mouth and throat if necessary.

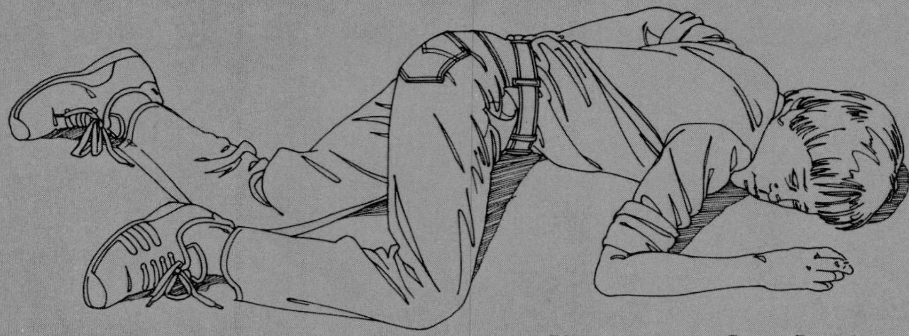

Place victim in Coma Position.

Coral Cuts

See page 53.

Infection after Marine Wounds

See pages 54 and 55.

Irritation from Spicules

May be due to

- sponges. See page 4.
- bristle worms. See page 63.
- sea squirts (ascidians). See page 122.

Itches

May be due to

- sponges. See page 4.
- schistosome dermatitis. See page 59.
- algal dermatitis. See page 60.
- bristle worms. See page 63.
- starfish. See page 106.
- sea cucumbers (beche-de-mer). See page 120.
- sea squirts (ascidians). See page 122.

Poisoning – from Eating

Symptoms range from numbness, weakness, drowsiness, a burning sensation in mouth and/or stomach to vomiting and diarrhoea, and even to breathing failure because of paralysis or air-passage obstruction after vomiting. Symptoms occur some time after eating animals containing paralytic shellfish poison (PSP) such as

- poisonous crabs. See pages 67, 68 and also pages 94, 95.
- molluscs affected by a red tide. See pages 99 to 101.
- venomous sea urchins. See page 118.

First Aid

1. Clear airways if blocked after vomiting. Beware of vomiting again, place victim on one side. See page 86.
2. Reassure victim, who may be speechless and paralysed.
3. Seek medical attention *but do not leave victim alone.*
4. Keep alive. Use artificial respiration if breathing fails. Use external cardiac massage if heart fades.

Poisoning – from Touching

Spines of some sea urchins and pincer organs of others inject venom causing severe pain, swelling, skin discoloration and sometimes paralysis.

This may be due to

- long spined sea urchins. See page 109.
- short spined sea urchins. See page 114.
- pincer organs of sea urchins. See page 116.

Stings – Mild to Moderate

Stings with pain, swelling and rash may be due to any animals with stinging capsules (nematocysts). See page 9.

- fireweeds (stinging hydroids). See page 11.
- portuguese man-of-war. See pages 15 and 16.
- fire coral (Millepora). See page 18.
- stinger, small box jellyfish *(Carybdea rastoni).* See page 21.
- mauve stinger (Pelagia). See pages 37 and 38.
- sea nettle (Chrysaora). See pages 38 and 39.
- hairy stinger (Cyanea). See pages 39 to 41.
- sea anemone. See pages 46 and 48.
- corals. See page 53.

First Aid

1. Apply vinegar liberally to inactivate the stinging capsules.
2. Remove adhering pieces of animal.
3. I.C.E. Immobilise, elevate and cool injured area with ice, cold cans and bottles etc.
4. Apply local anaesthetic cream.
5. Treat for general pain relief. See 'Severe Stings' below.
6. Refer to appropriate page given above for further information on treatment.

Stings – Severe

Stings with or without weals but with severe pain and shock may be due to

- large box jellyfish with multiple tentacles *Chironex* (sea wasp) and *Chiropsalmus.* See pages 32 and 33.
- large box jellyfish with four tentacles *Carybdea alata grandis.* See page 24. *Tamoya* sp. See page 25.
- small but dangerous jellyfish *Carukia* (Irukandji stinger). See page 23.

First Aid

1. Apply vinegar liberally.
2. Aim to keep victim alive until the short-acting venom wears off.
3. Use artificial respiration if breathing fails.
4. Use external cardiac massage if heartbeat begins to fade.
5. See medical attention for treatment for shock and pain *but do not leave victim alone.*
6. Use Box Jellyfish Antivenom for *Chironex* and *Chiropsalmus* if available.
7. Watch for allergic reactions which may also need urgent medical assistance.
8. For delayed reaction after a minor sting see page 23 *(Carukia).*

Wounds from Spines

May be due to

- crustaceans such as crabs or rock lobsters. See Fish-handlers Disease, page 55.
- crown-of-thorns starfish. See page 106.
- sea urchins. See pages 109, 111 and 114.

If wound becomes infected see pages 54 and 55.

Index

Numbers in bold type refer to pages which include figures